BUILD YOUR OWN
SOLAR WATER HEATER

BUILD YOUR OWN
SOLAR WATER HEATER

Stu Campbell

with Douglas Taff, Ph.D.

Illustrations by Robert Vogel

GARDEN WAY PUBLISHING
CHARLOTTE, VERMONT 05445

Printed in the United States
First printing, March 1978

Library of Congress Cataloging in Publication Data

Campbell, Stu, 1942 —
  Build your own solar water heater.

  Bibliography: p.
  Includes index.
  1. Solar heating.   2. Water heaters.
I. Taff, Douglas, joint author. II. Title.
TH7413.C35   696'.6   77-28692
ISBN 0-88266-129-9
ISBN 0-88266-128-0 pbk.

Contents

Foreword

The whole solar energy field is growing by astonishing — and sometimes confusing — leaps and bounds. But we know this much for sure: that *using the sun to heat domestic hot water is the most practical and least expensive application of solar energy at this time.* Heating water with the sun is incredibly simple; and although some experts try to make it seem otherwise, most of the techniques can be understood by any 7th grader.

Besides, solar water heating isn't new. It's been going on for centuries. There were some pretty elaborate and sophisticated solar heaters around by the end of the 19th century, and in 1951 there were as many as 50,000 solar water heaters in Miami, Florida alone. People in Japan, Israel and Australia have been relying on them for years. So there's a good body of knowledge and experience to draw from.

Build Your Own Solar Water Heater is strictly about solar water heating. Bruce Anderson, a respected solar expert at Total Environmental Action in New Hampshire says, "I strongly believe people have to know as much as possible about a solar system before putting any money into it." He's right, and I hope this book helps fill that bill.

What you should know from the start is that, yes, eventually solar power *will* be cheaper than fossil fuel, electricity, and probably even nuclear energy. After all, the most attractive thing about solar energy is that it's free. *But* the equipment needed to collect and store it is fairly expensive. And in spite of what a lot of people are saying, solar systems of all kinds are *not* likely to get cheaper. Granted, some manufacturing costs may go down, but the inflationary spiral will surely neutralize any possible savings.

The sooner you invest in a solar energy system, the earlier you'll be able to pay for it in fuel-expense savings. And once you've paid it off, except for incidental maintenance costs, it's all gravy from then on.

The second thing you should know before you begin is that everything in this book *works.* It's not just theory. We've built it, or we've done it, or we've tested it. And we've made plenty of mistakes — ones you can avoid. We've also had a lot of success and have a pretty good idea of how much it all should cost you.

Besides doing our own experimenting, we've explored what others have to say. We've been watching big industry as it begins to move into the solar field, and we've been keeping an eye on some of the "counter-culture" folks who have been there for some time, and whose funky inventive work is often more promising than what's coming out of the big companies.

All of the exploration boils down to this: If you're willing to do a lot of the installation work yourself, heating most of your hot water by the sun can be *very* economical. Even if you buy prefabricated components and hire a contractor to put them in, it may take somewhat longer to get your money back, but it will still be worthwhile in the long run.

We'll offer easy-to-understand suggestions about very simple hot water systems at first, and gradually go on to describe more complex ones. But all of the designs will have three fundamental components: (1) the *solar collector* itself, which transforms the sun's rays into usable heat, (2) the *transfer and storage system,* which moves the heat from the collector to a place where it can be kept and protected for later use, and (3) a *control system,* either automatic or manual, which regulates the whole heater — turning it on or off, speeding it up or slowing it down as the intensity of the sun and household demands dictate.

You can pick and choose as you please, confident that if you understand the easy principles involved, and can adapt them to your home's structure, you're sure to end up with a solar water heater that not only works well and saves you money, but makes you feel as though you're somehow beating the system. That's the most fun of all.

"There's no such thing as a free lunch," or so they say, but there *can* be free showers and baths. Stay in as long as you like, and don't worry about the gas, oil, or electric bill. Enjoy the fruits of a sunny day — even after dark. Solar hot water can cost little or nothing.

And when you step into the shower, remember the remark that appears in the bottom corner of Steve Baer's Zomeworks blueprint for a simple solar hot water heater. He says, "Solar hot water gets you cleaner, makes you happier, rests your bones, cures impotence. If you don't believe it, ask someone who uses it."

Maybe we won't go *that* far, but we can say this much: once your solar water heater is working, it's likely to become a curiosity, and you may become a local celebrity. People may want to visit and have a look. Be prepared for lots of questions and lots of small talk if you let them come. Whether or not you decide to invite them into the shower is entirely up to you.

STU CAMPBELL

To my Grandfather

Who knew a lot about this stuff long before it became a necessity.

Special thanks to Art Brunig, master plumber of Stowe, Vermont for his patient explanations and help.

CHAPTER 1

Dollars and Sense

Unless you're very rich — or very trusting — you'll want to sit down with a pencil and paper and do some "ciphering" before you decide if and how you're going to put a solar hot water system in your home.

Various types of solar water heaters are at work right now, in all parts of the United States, saving people money. That's a pretty safe statement. But don't accept it on faith. If you can make some simple calculations you can get a fairly accurate picture of how much free hot water you can expect and how long it will take to get back your initial investment. Let's get some of the numbers out of the way before getting down to other details.

Try to keep in mind that there are all kinds of variables in these figures. To keep ourselves honest, let's use very conservative efficiency figures and very *high* cost estimates for a theoretical hot water system that's fairly complex (see Figure 1). In other words, because we'll be using figures that are extreme, you may be able to do better than what you see here, just by building a simpler system, by finding cheaper parts, or by making your heater more efficient.

How Much Hot Water Do We Use?

Americans use — and waste — appalling amounts of hot water. People in other parts of the world just can't believe us. Motel owners in

Europe complain bitterly about how tourists from the United States increase their fuel bills and overload their septic systems. We're spoiled. We squander water just the way we've squandered many of our other resources.

John Murphy, in the *Homeowner's Energy Guide,* says that we use up hot water in approximately this way: Every time we take a shower we use about 10 gallons of hot water. Whenever we take a bath we use at least that much — often more. An average clothes washer uses 22 gallons of hot water per regular washing cycle, and an automatic dishwasher uses about 15 gallons — which, by the way, can be *more* hot water than it takes to wash dishes in the sink by hand.

Each time we shave or wash our hands we use about 1/2 gallon of hot water, on the average. But what's really shocking is how much we waste just testing the temperature of our shower water — 1/2 to 2 gallons a test! We're literally pouring energy down the drain. (See Panel 1.)

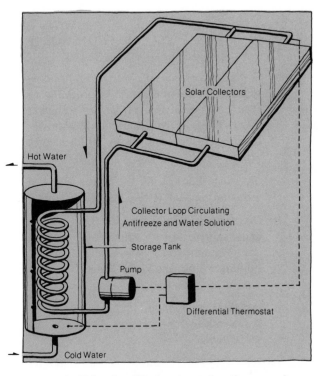

Figure 1. This simplified schematic diagram shows the three major components in a solar-operated hot water system: (1) the collectors themselves, (2) the pump, plumbing, storage tank and heat-exchanger coil, and (3) the differential thermostat that controls the whole system.

All of this averages out to 2,500 gallons of hot water a month for a family of 5, or 30,000 gallons a year. That may be twice as much as we actually need, but it's what we're used to. This *can* be reduced of course, in any number of ways: by lowering the thermostat on the hot water heater, insulating hot water tanks and pipes, using cold water detergents and washing clothes in cold water, and — best of all — taking a shower with a friend.

How Much Does Hot Water Cost Now?

When you get a conventional hot water heater from the factory, the thermostat is normally pre-set at 165 degrees Fahrenheit. That's a lot hotter than the water needs to be. Human skin can comfortably stand water temperatures in the 105 to 115 degree range, and this is about the temp-erature of the water that comes out of your shower head.

To get water to this temperature we usually mix hot and cold water together, which should be enough to convince us that there's really no need for water that's as high as 165 degrees. We'd do just fine to turn the heater down to as low as 130°. (Below this setting we would have to run less cold water with hot, but we would run out of hot water much sooner.)

Yet all indications are that most people don't — partly because they don't know they should, and partly because thermostats, particularly ones on older electric water heaters, are hard to find, much less get at. Because this is the case, let's use 165 degrees as a point of departure so we can do some arithmetic.

PANEL 1

WAYS FAMILIES CAN CONTROL HOT WATER USAGE

1. *Showers instead of baths.* Baths nor-mally use more and hotter water. Showers should be short. Set a timer for a shower taker. If you install a water-saving shower head, you'll be using up to 40% less hot water.

2. *Shaving.* Partially fill the basin with hot water rather than letting the faucet run.

3. *Dishwashing by hand.* It may not be necessary to wash dishes three times a day. Washing once will save hot water, human energy, and time. Rinse dishes in a pan rather than under running water.

4. *Dishwasher.* Wash full loads. Don't handwash before putting dishes in the dishwasher.

5. *Food disposal.* Always use cold water.

6. *Leaky faucets.* Fix them right away. One dripping hot water faucet can cost up to $15 extra per year if not repaired.

7. *Laundry.* Wash full loads. Wash in hot water, but rinse in cold water.

8. *Stagger use of hot water.* Eliminate heavy demand on your supply of hot water in a short period of time.

9. *Water heater.* Shutting off the "quick-recovery" element in an electric hot water heater will not save energy. Instead, turn the thermostat down. If away from home for extended periods of time, turn it down still further. Precautions should be taken to prevent pipes from freezing in winter.

10. *Insulation.* Insulate the hot water heater with an extra 6″ of fiberglass. This much insulation should pay for itself in just 2 to 3 months.

Parts borrowed from Cooperative Extension Service, Pennsylvania State University.

The minute we start thinking about heat, we have to understand what a BTU is. "BTU" stands for British Thermal Unit, and a BTU is just a simple measuring unit for heat. One BTU is the amount of heat needed to raise the temperature of one pound of water one degree Fahrenheit. You may already know that. The problem comes when we have to convert pounds of water to gallons — because we think of water in terms of gallons, not pounds. A gallon of water weighs 8.33 pounds, which means that it takes 8.33 BTU's to heat a gallon of water one degree.

Now, the normal temperature of ground water in most parts of the world is around 55 degrees Fahrenheit, give or take a couple of degrees. Water from a well, spring, or even a municipal water main will probably be this cold when it comes into your house.

If we want to raise the temperature of this water to 165 degrees Fahrenheit, we'll have to increase its temperature by 110 degrees (165 minus 55). If we then multiply 110 degrees temperature difference by 8.33 pounds — the weight of one gallon of water — we come up with a figure of 916.30.

$$110 \times 8.33 = 916.30 \text{ BTU's}$$

This means that it takes over 900 BTU's to heat a single gallon of water to the temperature that the preset thermostat on the hot water heater is asking for.

Over the course of a year the 30,000 gallons of hot water that our imaginary family of 5 uses would require 27,489,000 BTU's to heat (916.30 × 30,000). Of course some of this heat is lost. It leaks out of the hot water tank and sneaks out of the plumbing every chance it gets on its way to the shower or dishwasher. So the 27.5 million BTU figure can't be accurate. It's pretty hard to say how much heat is actually lost, but estimating conservatively, we can safely guess that 2.5 million BTU's are lost in a year's time. So the real heat requirement looks more like 30 million BTU's a year (Panel 2).

Heating costs are most easily measured in terms of dollars per million BTU's. A million BTU's of electricity costs about $14.64 on the average in 1977. (That's figuring electricity at 5¢ a

PANEL 2

1975 ELECTRICAL RATES IN SELECTED CITIES

1. New York, N.Y.	$0.089 per kwh
2. Springfield, Mass.	0.062 per kwh
3. Boston, Mass.	0.061 per kwh
4. Rutland, Vt.	0.059 per kwh
5. Pasadena, Calif.	0.055 per kwh
6. Philadelphia, Pa.	0.053 per kwh
7. Providence, R.I.	0.052 per kwh
8. Tucson, Ariz.	0.051 per kwh
9. Lebanon, Pa.	0.050 per kwh
10. New Haven, Conn.	0.049 per kwh
11. Los Angeles, Calif.	0.048 per kwh
12. Waterloo, Iowa	0.047 per kwh
13. Baltimore, Md.	0.046 per kwh
14. Chicago, Ill.	0.046 per kwh
15. St. Petersburg, Fla.	0.045 per kwh
16. Springfield, Mo.	0.044 per kwh
17. Tampa, Fla.	0.044 per kwh
18. El Paso, Texas	0.043 per kwh
19. Flint, Mich.	0.042 per kwh
20. Southhampton, N.Y.	0.041 per kwh
1. Detroit, Mich.	0.041 per kwh
2. New Orleans, La.	0.040 per kwh
3. San Antonio, Texas	0.040 per kwh
4. Troy, N.Y.	0.040 per kwh
5. Greenwich, Conn.	0.039 per kwh
6. Minneapolis, Minn.	0.039 per kwh
7. Hartford, Conn.	0.039 per kwh
8. Milwaukee, Wis.	0.039 per kwh
9. Denver, Colo.	0.038 per kwh
10. Little Rock, Ark.	0.038 per kwh
11. Columbia, S.C.	0.037 per kwh
12. Roanoke, Va.	0.037 per kwh
13. Dayton, Ohio	0.036 per kwh
14. Wheeling, W. Va.	0.035 per kwh
15. Cincinnati, Ohio	0.034 per kwh
16. San Francisco, Calif.	0.033 per kwh
17. Raleigh, N.C.	0.032 per kwh
18. Lincoln, Neb.	0.031 per kwh
19. Birmingham, Ala.	0.030 per kwh
20. Sheridan, Wyo.	0.029 per kwh

From Sept. 1975 AHAM Directory.

kilowatt hour.) One million BTU's of fuel oil — at 50¢ a gallon — will cost somewhere around $5.10 (at normal 70% burner efficiency), and a million BTU's of natural gas will cost $4.00 at 34¢ a cubic foot at normal burner efficiency. If we multiply these costs by 30 — because we're talking about 30,000,000 BTU's — here's how things look for a whole year:*

Temperature Setting	Electricity	Fuel Oil	Natural Gas
1. 165 degrees F.	$439.20	$153.00	$120.00
2. 160 degrees F.	$423.09	$147.39	$115.60
3. 150 degrees F.	$383.56	$133.62	$104.80
4. 140 degrees F.	$348.43	$121.38	$ 95.20
5. 130 degrees F.	$311.10	$108.37	$ 86.00
6. 120 degrees F.	$278.16	$ 96.90	$ 76.00

If we can agree that 165 degrees is an unreasonably high setting for a hot water heater, let's take a look at line 4 in the table shown here. This is the 140 degree setting — which may make more sense.

These figures represent somewhere between 15 and 25% of the family's total energy dollar. Taking it one giant step further, about 4% of the nation's energy consumption goes into heating domestic hot water (see Panel 3).

How Much Energy Do You Need to Heat Your Water?

Here's how to compute how much of an energy load you put on your hot water system now:

$$Q = G \times K \times \Delta T$$

This standard formula looks pretty complicated, but it's simple once you understand what all the symbols stand for.

Your family's BTU requirements for a day (Q) equals the number of gallons of hot water used in

---

* Above 140° F. cold water is normally mixed with hot — unless you're doing laundry. So the cost figures in lines 1, 2 and 3 will seem quite high.

PANEL 3

ENERGY USAGE IN AMERICAN HOMES

Heating of space	57.5%
Water heating	14.9%
Refrigeration	6.0%
Cooking	5.5%
Air conditioning	3.7%
Lighting	3.5%
Television	3.0%
Food freezer	1.9%
Clothes dryer	1.7%
Other	2.3%

U.S. Department of Agriculture (1974)

a day (G) multiplied by the weight of water (8.33 pounds per gallon — or K) times the difference between your incoming water supply and the chosen water temperature at the faucet (ΔT).

The average American individual uses about 20 gallons of hot water each day. The amount of water used in a household in one day can be easily figured by multiplying the number of people in the family by 20 gallons. Let's stick with a family of 5, since that's where we started. So a family of 5 uses 100 gallons of hot water every day.

PANEL 4

BTU REQUIREMENTS FOR HEATING DOMESTIC HOT WATER AT 140 DEGREES F.

Number of persons	BTU's Needed
1	14,161/day
2	28,322/day
3	42,483/day
4	56,644/day
5	70,805/day
6	84,966/day
7	99,127/day
8	113,288/day

We have already figured out what ΔT is all about. If we subtract the temperature of the water coming into the heater (55 degrees) from the temperature of the heated water (140 degrees) we get the temperature differential — or ΔT — which is 85 degrees. So our equation looks like this:

$$Q = 100 \times 8.33 \times 85 = 70,805 \text{ BTU's per day}$$

Again, what all this means is that for every 100 gallons of hot water that you and yours use each day, your present system has to generate almost 71,000 BTU's of heat. A system for a family of 4 will have to generate 56,664 BTU's per day (Panel 4). What you now need to figure out is what percentage of your total BTU requirements you will be able to generate with solar power (Tables 1 and 2).

Table 1. This table gives you an idea of how many BTU's of solar energy fall on a square foot of surface in one hour. Note how the numbers vary from month to month and from one latitude to the next. On a horizontal surface (0 degrees) the smallest amount of solar energy falls in December, and the greatest amount in June. But as the steepness of the roof increases toward vertical (90 degrees) the figures change. Because solar collectors are traditionally mounted on rooftops, the angle of a south-facing roof is an important consideration.

TABLE 1

DAILY SUNSHINE ON A SOUTH-FACING ROOF

24° N. LATITUDE
BTU/Hr.-Sq. Ft. on South Facing Surface
Angle with Horizontal

Date	0°	14°	24°	34°	54°	90°
Sept. 21	2194	2342	2366	2322	2212	992
Oct. 21	1928	2198	2314	2364	2346	1442
Nov. 21	1610	1962	2146	2268	2324	1730
Dec. 21	1474	1852	2058	2204	2286	1808
Jan. 21	1622	1984	2174	2300	2360	1766
Feb. 21	1998	2276	2396	2446	2424	1476
Mar. 21	2270	2428	2456	2412	2298	1022
Apr. 21	2454	2458	2374	2228	2016	488
May 21	2556	2447	2286	2072	1800	246
June 21	2574	2422	2230	1992	1700	204
July 21	2526	2412	2250	2036	1766	246
Aug. 21	2408	2402	2316	2168	1958	470

32° N. LATITUDE

Date	0°	22°	32°	42°	52°	90°
Sept. 21	2014	2288	2308	2264	2154	1226
Oct. 21	1654	2100	2208	2252	2232	1588
Nov. 21	1280	1816	1980	2084	2130	1742
Dec. 21	1136	1704	1888	2016	2086	1794
Jan. 21	1288	1839	2008	2118	2166	1779
Feb. 21	1724	2188	2300	2345	2322	1644
Mar. 21	2084	2378	2403	2358	2246	1276
Apr. 21	2390	2444	2356	2206	1994	764
May 21	2582	2454	2284	2064	1788	469
June 21	2634	2436	2234	1990	1690	370
July 21	2558	2422	2250	2030	1754	458
Aug. 21	2352	2388	2296	2144	1934	736

40° N. LATITUDE

Date	0°	30°	40°	50°	60°	90°
Sept. 21	1788	2210	2228	2182	2074	1416
Oct. 21	1348	1962	2060	2098	2074	1654
Nov. 21	942	1636	1778	1870	1908	1686
Dec. 21	782	1480	1634	1740	1796	1646
Jan. 21	948	1660	1810	1906	1944	1726
Feb. 21	1414	2060	2162	2202	2176	1730
Mar. 21	1852	2308	2330	2284	2174	1484
Apr. 21	2274	2412	2320	2168	1956	1022
May 21	2552	2442	2264	2040	1760	724
June 21	2648	2434	2224	1974	1670	610
July 21	2534	2409	2230	2006	1728	702
Aug. 21	2244	2354	2258	2104	1894	978

48° N. LATITUDE

Date	0°	38°	48°	58°	68°	90°
Sept. 21	1522	2102	2118	2070	1966	1546
Oct. 21	1022	1774	1860	1890	1866	1626
Nov. 21	596	1336	1448	1518	1544	1442
Dec. 21	446	1136	1250	1326	1364	1304
Jan. 21	596	1360	1478	1550	1578	1478
Feb. 21	1080	1880	1972	2024	1978	1720
Mar. 21	1578	2208	2228	2182	2074	1632
Apr. 21	2106	2358	2266	2114	1902	1262
May 21	2482	2418	2234	2010	1728	982
June 21	2626	2420	2204	1950	1644	874
July 21	2474	2386	2200	1974	1694	956
Aug. 21	2086	2300	2200	2046	1836	1208

After ASHRAE Data 1973

TABLE 2

MEAN PERCENTAGE OF POSSIBLE SUNSHINE FOR SELECTED LOCATIONS

Table 2. The sun only shines a certain percentage of the time — mostly because of cloud cover. The figures in Table 2 give you a month-by-month reading of the percentage of possible sunshine in different cities across the U.S. In most cases the month with the least sunshine is January. This is when it becomes most difficult to generate enough heat from the sun to heat a large percentage of your domestic hot water.

MEAN PERCENTAGE OF POSSIBLE SUNSHINE FOR SELECTED LOCATIONS

STATE AND STATION	YEARS	JAN.	FEB.	MAR.	APR.	MAY	JUNE	JULY	AUG.	SEPT.	OCT.	NOV.	DEC.	ANNUAL
ALA. BIRMINGHAM	56	43	49	56	63	66	67	62	65	66	67	58	44	59
MONTGOMERY	49	51	53	61	69	73	72	66	69	69	71	64	48	64
ALASKA, ANCHORAGE	19	39	46	56	58	50	51	45	39	35	32	33	29	45
FAIRBANKS	20	34	50	61	68	55	53	45	35	31	28	38	29	44
JUNEAU	14	30	32	39	37	34	35	28	30	25	18	21	18	30
NOME	29	44	46	48	53	51	48	32	26	34	35	36	30	41
ARIZ. PHOENIX	64	76	79	83	88	93	94	84	84	89	88	84	77	85
YUMA	52	83	87	91	94	97	98	92	91	93	93	90	83	91
ARK. LITTLE ROCK	66	44	53	57	62	67	72	71	73	71	74	58	47	62
CALIF. EUREKA	49	40	44	50	53	54	56	51	46	52	48	42	39	49
FRESNO	55	46	63	72	83	89	94	97	97	93	87	73	47	78
LOS ANGELES	63	70	69	70	67	68	69	80	81	80	76	79	72	73
RED BLUFF	39	50	60	65	75	79	86	95	94	89	77	64	50	75
SACRAMENTO	48	44	57	67	76	82	90	96	95	92	82	65	44	77
SAN DIEGO	68	68	67	68	66	60	60	67	70	70	70	76	71	68
SAN FRANCISCO	64	53	57	63	69	70	75	68	63	70	70	62	54	66
COLO. DENVER	64	67	67	65	63	61	69	68	68	71	71	67	65	67
GRAND JUNCTION	57	58	62	64	67	71	79	76	72	77	74	67	58	69
CONN. HARTFORD	48	46	55	56	54	57	60	62	60	57	55	46	46	56
D. C. WASHINGTON	66	46	53	56	57	61	64	64	62	62	61	54	47	58
FLA. APALACHICOLA	26	59	62	62	71	77	70	64	63	62	74	66	53	65
JACKSONVILLE	60	58	59	66	71	71	63	62	63	58	58	61	53	62
KEY WEST	45	68	75	78	78	76	70	69	71	65	65	69	66	71
MIAMI BEACH	48	66	72	73	73	68	62	65	67	62	62	65	65	67
TAMPA	63	63	67	71	74	75	66	61	64	64	67	67	61	68
GA. ATLANTA	65	48	53	57	65	68	68	62	63	65	67	60	47	60
HAWAII. HILO	9	48	42	41	34	31	41	44	38	42	41	34	36	39
HONOLULU	53	62	64	60	62	64	66	67	70	70	68	63	60	65
LIHUE	9	48	48	48	46	51	60	58	59	67	58	51	49	54
IDAHO, BOISE	20	40	48	59	67	68	75	89	86	81	66	46	37	66
POCATELLO	21	37	47	58	64	66	72	82	81	78	66	48	36	64
ILL. CAIRO	30	46	53	59	65	71	77	82	79	75	73	56	46	65
CHICAGO	66	44	49	53	56	63	69	73	70	65	61	47	41	59
SPRINGFIELD	59	47	51	54	58	64	69	76	72	73	64	53	45	60
IND. EVANSVILLE	48	42	49	55	61	67	73	78	76	73	67	52	42	64
FT. WAYNE	48	38	44	51	55	62	69	74	69	64	58	41	38	57
INDIANAPOLIS	63	41	47	49	55	62	68	74	70	68	64	48	39	59
IOWA. DES MOINES	66	56	56	56	59	62	66	75	70	64	64	53	48	62
DUBUQUE	54	48	52	52	58	60	63	73	67	61	55	44	40	57
SIOUX CITY	52	55	58	58	59	63	67	75	72	67	65	53	50	63
KANS. CONCORDIA	52	60	60	62	63	65	73	79	76	72	70	64	58	67
DODGE CITY	70	67	66	68	68	68	74	78	78	76	75	70	67	71
WICHITA	46	61	63	64	64	66	73	80	77	73	69	67	59	69
KY. LOUISVILLE	59	41	47	52	57	64	68	72	69	68	64	51	39	59
LA. NEW ORLEANS	69	49	50	57	63	66	64	58	60	64	70	60	46	59
SHREVEPORT	18	48	54	58	60	69	78	79	80	79	77	65	60	69
MAINE, EASTPORT	58	45	51	52	52	51	53	55	57	54	50	37	40	50
MASS. BOSTON	67	47	56	57	56	59	62	64	63	61	58	48	48	57
MICH. ALPENA	45	29	43	52	56	59	64	70	64	52	44	24	22	51
DETROIT	69	34	42	48	52	58	65	69	66	61	54	35	29	53
GRAND RAPIDS	56	26	37	48	54	60	66	72	67	58	50	31	22	49
MARQUETTE	55	31	40	47	52	53	56	63	57	47	38	24	24	47
S. STE. MARIE	60	28	44	50	54	54	59	63	58	45	36	21	22	47
MINN. DULUTH	49	47	55	60	58	58	60	68	63	53	47	36	40	55
MINNEAPOLIS	45	49	54	55	57	60	64	72	69	60	54	40	40	56
MISS. VICKSBURG	66	46	50	57	64	69	73	69	72	74	71	60	45	64
MO. KANSAS CITY	69	55	57	59	60	64	70	76	73	70	67	59	52	65
ST. LOUIS	68	48	49	56	59	64	68	72	68	67	65	54	44	61
SPRINGFIELD	45	48	54	57	60	63	69	77	72	71	65	58	48	63
MONT. HAVRE	55	49	58	61	63	63	65	78	75	64	57	48	46	62
HELENA	65	46	55	58	60	59	63	77	74	63	57	48	43	60
KALISPELL	50	28	40	49	57	58	60	77	73	61	50	28	20	53
NEBR. LINCOLN	55	57	59	60	60	63	69	76	71	67	66	59	55	64
NORTH PLATTE	53	63	63	64	62	64	72	78	74	72	70	62	58	68
NEV. ELY	21	61	64	68	65	67	79	79	81	81	73	67	62	72
LAS VEGAS	19	74	77	78	81	85	91	84	86	92	84	83	75	82
RENO	51	59	64	69	75	77	82	90	89	86	76	68	56	76
WINNEMUCCA	53	52	60	64	70	76	83	90	90	86	75	62	53	74
N. H. CONCORD	44	48	53	55	53	51	56	57	58	55	50	43	43	52
N. J. ATLANTIC CITY	62	51	57	58	59	62	65	67	66	65	54	58	52	60

MEAN PERCENTAGE OF POSSIBLE SUNSHINE FOR SELECTED LOCATIONS

STATE AND STATION	YEARS	JAN.	FEB.	MAR.	APR.	MAY	JUNE	JULY	AUG.	SEPT.	OCT.	NOV.	DEC.	ANNUAL
N. MEX. ALBUQUERQUE	28	70	72	72	76	79	84	76	75	81	80	79	70	76
ROSWELL	47	69	72	75	77	76	80	76	75	74	74	74	69	74
N. Y. ALBANY	63	43	51	53	53	57	62	63	61	58	54	39	38	53
BINGHAMTON	63	31	39	41	44	50	56	54	51	47	43	29	26	44
BUFFALO	49	32	41	49	51	59	67	70	67	60	51	31	28	53
CANTON	43	37	47	50	48	54	61	63	61	54	45	30	31	49
NEW YORK	83	49	56	57	59	62	65	66	64	64	61	53	50	59
SYRACUSE	49	31	38	45	50	58	64	67	63	56	47	29	26	50
N. C. ASHEVILLE	57	48	53	56	61	64	63	59	59	62	64	59	48	58
RALEIGH	61	50	56	59	64	67	65	62	62	63	64	62	52	61
N. DAK. BISMARCK	65	52	58	56	57	58	61	73	69	62	59	49	48	59
DEVILS LAKE	55	53	60	59	60	59	62	71	67	59	56	44	45	58
FARGO	39	47	55	56	58	62	63	73	69	60	57	39	46	59
WILLISTON	43	51	59	60	63	66	66	78	75	65	60	48	48	63
OHIO, CINCINNATI	44	41	46	52	56	62	69	72	68	68	60	46	39	57
CLEVELAND	65	29	36	45	52	61	67	71	68	62	54	32	25	50
COLUMBUS	65	36	44	49	54	63	68	71	68	66	60	44	35	55
OKLA, OKLAHOMA CITY	62	57	60	63	64	65	74	78	78	74	68	64	57	68
OREG. BAKER	46	41	49	56	61	63	67	83	81	74	62	46	37	60
PORTLAND	69	27	34	41	49	52	55	70	65	55	42	28	23	48
ROSEBURG	29	24	32	40	51	57	59	79	77	68	42	28	18	51
PA. HARRISBURG	60	43	52	55	57	61	65	68	63	62	58	47	43	57
PHILADELPHIA	66	45	56	57	58	61	62	64	61	62	61	53	49	57
PITTSBURGH	63	32	39	45	50	57	62	64	61	62	54	39	30	51
R. I. BLOCK ISLAND	48	45	54	47	56	58	60	62	62	60	59	50	44	56
S. C. CHARLESTON	61	58	60	65	72	73	70	66	66	67	68	68	57	66
COLUMBIA	55	53	57	62	68	69	68	63	65	64	68	64	51	63
S. DAK. HURON	62	55	62	60	62	65	68	76	72	66	61	52	49	63
RAPID CITY	53	58	62	63	62	61	66	73	73	69	66	58	54	64
TENN. KNOXVILLE	62	42	49	53	59	64	66	64	59	64	64	53	41	57
MEMPHIS	55	44	51	57	64	68	74	73	74	70	69	58	45	64
NASHVILLEE	63	42	47	54	60	65	69	69	68	69	65	55	42	59
TEX. ABILENE	14	64	68	73	66	73	86	83	85	73	71	72	66	73
AMARILLO	54	71	71	75	75	75	82	81	81	79	76	76	70	76
AUSTIN	33	46	50	57	60	62	72	76	79	70	70	57	49	63
BROWNSVILLE	37	44	49	51	57	65	73	78	78	67	70	54	44	61
DEL RIO	36	53	55	61	63	60	66	75	80	69	66	58	52	63
EL PASO	53	74	77	81	85	87	87	78	78	80	82	80	73	80
FT. WORTH	33	56	57	65	66	67	75	78	78	74	70	63	58	68
GALVESTON	66	50	50	55	61	69	76	72	71	70	74	62	49	63
SAN ANTONIO	57	48	51	56	58	60	69	74	75	69	67	55	49	62
UTAH, SALT LAKE CITY	22	48	53	61	68	73	78	82	82	84	73	56	49	69
VT. BURLINGTON	54	34	43	48	47	53	59	62	59	51	43	25	24	46
VA. NORFOLK	60	50	57	60	63	67	66	66	66	63	64	60	51	62
RICHMOND	56	49	55	59	63	67	66	65	62	63	64	58	50	61
WASH. NORTH HEAD	44	28	37	42	48	48	48	50	46	48	41	31	27	41
SEATTLE	26	27	34	42	48	53	48	62	56	53	36	28	24	45
SPOKANE	62	26	41	53	63	64	68	82	79	68	53	28	22	58
TATOOSH ISLAND	49	26	36	39	45	47	46	48	44	47	38	26	23	40
WALLA WALLA	44	24	35	51	63	67	72	86	84	72	59	33	20	60
YAKIMA	18	34	49	62	70	72	74	86	86	74	61	38	29	65
W. VA. ELKINS	55	33	37	42	47	55	55	56	53	55	51	41	33	48
PARKERSBURG	62	30	36	42	49	56	60	63	60	60	53	37	29	48
WIS. GREEN BAY	57	44	51	55	56	58	64	70	65	58	52	40	40	55
MADISON	59	44	49	52	53	58	64	70	66	60	56	41	38	56
MILWAUKEE	59	44	48	53	56	60	65	73	67	62	56	44	39	57
WYO. CHEYENNE	63	65	66	64	61	59	68	70	68	69	69	65	63	66
LANDER	57	66	70	71	66	65	74	76	75	72	67	61	62	69
SHERIDAN	52	56	61	62	61	61	67	76	74	67	60	53	52	64
YELLOWSTONE PARK	35	39	51	55	57	56	63	73	71	65	57	45	38	56
P. R. SAN JUAN	57	64	69	71	66	59	62	65	67	61	63	63	65	65

Here you need to stop and be very honest with yourself. Unless you live near the equator — where the sun is very strong all year long — it's unrealistic to even think about trying to make your solar water system 100% free of utility bills. There will be certain times during the year when you just can not collect or store that many BTU's from the sun, and will have to either rely on a conventional back-up system or go without hot water altogether.

But it *is* realistic to think about being between 50 and 90% independent of utility costs. In fact, 75% efficiency might be a reasonable target to shoot for. Don't forget, once your system has finished paying for itself, you would be getting 75% of your hot water for nothing. It also means that in the summer, and part of the spring and fall, you are getting 100% of your heated water for free, because 75% represents the yearly average. Statistics show, by the way, that we don't use any less hot water in the summer than we do during the colder months, so that's something else to keep in mind.

Without getting into a lot of heavy mathematics at this time, we can say this much: It should be easy to get 50% efficiency with any reasonably designed solar system. But after that you get diminishing returns. It gets harder and harder to achieve higher efficiencies — especially during December, January and February here in the

northern hemisphere. And it will be almost impossible to get 100% during these months, even with a whole roof full of solar collectors. But with some careful planning and a little know-how, 75% annual efficiency is well within the realm of possibility in most parts of the country (Figure 2).

How Many Collectors Will You Need?

This is a very complicated question. There are as many oversimplified rules of thumb for answering it, as there are unbelievably complex formulas that try to include all of the different factors involved. As you get further into the book, you will begin to see how intricate the variables become.

For now, look at Table 3. It's an attempt to give you a *rough* idea of how many collectors you will need where you are. It lists 80 cities in the United States and Canada, but chances are good that your home town does not appear there. You'll have to find the city closest to you and use it as your reference.

The table assumes that you have a roof on your house that faces south or close to south — within 20 degrees. If your roof faces more than 20 degrees away from south you may need to add up to 50% more collector area. If the ridge of your home runs north and south so roofs face east and west, you may need as much as twice the collector area recommended in the table. (There's no law that says solar collectors have to be mounted on a roof. There may be some place else that's more convenient.)

Notice as you study Table 3 that these are estimated summer and wintertime requirements for just *one* person. Look at Columbus, Ohio for example. A family of 4 there would need 4 times 0.7 or 2.8 collectors (67.2 square feet) in the summer, and 4 times 3.1 or 12.4 collectors (297.6 square feet) in the winter to satisfy 90 to 100% of their hot water load. You can average the two to get a reasonable estimate for year-round operation.

You should round off the number of collectors once you have multiplied the factor in the table

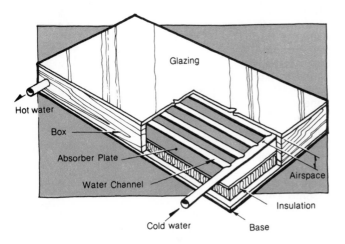

Figure 2. Cut-away sketch of a typical flat plate solar collector. Water running through the channels in the collector plate is heated by the sun and returned to the storage tank elsewhere. The glass on top traps radiation and holds heat inside the panel.

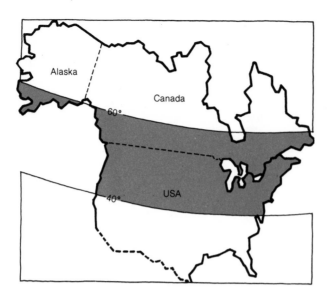

Table 3. The numbers in Table 3 are the result of some fairly complex calculations. The table tells you how many solar collector panels would be needed *per person* in different parts of the country in both summer and winter. The figures assume that a single solar collector has 24 square feet of surface area — although not all solar collectors are this size. This recommended number of collectors should satisfy 90 to 100% of one person's hot water needs.

TABLE 3

City	Approx. Latitude	Summer Collector* Requirement (per person)	Winter Collector Requirement
Albuquerque, NM	(35°)	0.5	1.4
Annette Island, AK	(55°)	1.2	4.0
Apalachicola, FL	(30°)	0.7	1.2
Astona, OR	(46°)	0.9	2.8
Atlanta, GA	(33°)	0.7	1.8
Barrow, AK	(71°)	1.3	175.4
Bethel, AK	(60°)	1.2	11.2
Bismarck, ND	(47°)	0.7	2.9
Blue Hill, MA	(42°)	0.9	2.9
Boise, ID	(43°)	0.6	2.9
Boston, MA	(42°)	0.8	2.9
Brownsville, TX	(26°)	0.6	1.2
Caribou, ME	(47°)	0.9	4.0
Charleston, SC	(33°)	0.7	1.5
Cleveland, OH	(41°)	0.8	3.2
Columbia, MO	(39°)	0.7	2.3
Columbus, OH	(40°)	0.7	3.1
Davis, CA	(38°)	0.6	2.1
Dodge City, KS	(38°)	0.6	1.6
East Lansing, MI	(42°)	0.8	3.7
East Wareham, MA	(42°)	0.9	2.4
Edmonton, Alberta	(53°)	0.9	4.6
El Paso, TX	(32°)	0.5	1.3
Ely, NV	(39°)	0.6	1.8
Fairbanks, AK	(65°)	0.9	83.3
Fort Worth, TX	(33°)	0.6	1.4
Fresno, CA	(37°)	0.6	1.9
Gainesville, FL	(29°)	0.7	1.2
Glasgow, MT	(48°)	0.6	3.4
Grand Junction, CO	(39°)	0.6	1.8
Grand Lake, CO	(40°)	0.8	2.6
Great Falls, MT	(47°)	0.7	2.8
Greensboro, NC	(36°)	0.7	1.9
Griffin, GA	(33°)	0.7	1.7
Hatteras, NC	(35°)	0.6	1.5
Indianapolis, IN	(40°)	0.7	2.9
Inyokem, CA	(35°)	0.5	1.1
Ithaca, NY	(42°)	0.8	3.6
Lake Charles, LA	(30°)	0.7	1.4
Lander, WY	(30°)	0.7	2.2
Las Vegas, NV	(36°)	0.5	1.2
Lemont, IL	(41°)	0.8	2.6
Lexington, KY	(38°)	0.7	2.0
Lincoln, NE	(41°)	0.8	2.1
Little Rock, AR	(34°)	0.7	1.8
Los Angeles, CA	(34°)	0.7	1.2
Madison, WI	(43°)	0.8	2.8
Matanuska, AK	(61°)	1.1	14.2
Medford, OR	(42°)	0.6	3.1
Miami, FL	(26°)	0.7	0.9
Midland, TX	(32°)	0.6	1.6
Nashville, TN	(36°)	0.7	2.1
Newport, RI	(41°)	0.8	2.4
New York, NY	(41°)	0.8	2.5
Oak Ridge, TN	(36°)	0.7	2.1
Oklahoma City, OK	(35°)	0.6	1.4
Ottawa, Ontario	(45°)	0.8	3.6
Phoenix, AZ	(33°)	0.5	1.2
Portland, ME	(43°)	0.8	2.3
Rapid City, SD	(44°)	0.7	2.2
Riverside, CA	(34°)	0.6	1.2
St. Cloud, MN	(45°)	0.7	3.4
Salt Lake City, UT	(41°)	no data	2.3
San Antonio, TX	(29°)	0.6	1.3
Santa Maria, CA	(35°)	0.7	1.4
Sault Ste. Marie, MI	(46°)	0.8	3.8
Sayville, NY	(40°)	0.8	2.4
Schenectady, NY	(43°)	0.9	3.7
Seattle, WA	(47°)	0.9	3.7
Seabrook, NJ	(39°)	0.8	2.4
Spokane, WA	(47°)	0.6	4.0
State College, PA	(41°)	0.8	3.1
Stillwater, OK	(36°)	0.7	1.6
Tampa, FL	(28°)	0.7	1.0
Toronto, Ontario	(43°)	0.8	3.6
Tucson, AZ	(32°)	0.6	1.1
Upton, NY	(41°)	0.8	2.2
Washington, DC	(39°)	0.7	2.1
Winnipeg, Manitoba	(50°)	0.8	5.0

*Collector assumes 24 sq. ft. of absorbing surface.

8

by the number of people in the family. If you average the summertime needs and the winter needs and end up with a figure of 3.3 collectors, plan on 3. If you calculate that you need 4.7 collector panels, go to 5, and so on.

Think about fewer collectors at first, rather than more. You can always add collectors later on if you feel you can increase your efficiency significantly.

Your costs could be astronomical if you try to have enough collectors to supply 90 to 100% of your wintertime domestic hot water needs. But you can be pretty well assured that the number of collectors needed to give you 90 to 100% efficiency in the summer will supply about 45 to 50% of your year-round needs, once everything is averaged out.

In other words, you're going to have to work out a compromise. You'll have to balance what your system can do in December — probably too little — against what it can do in June — probably more than enough. Whether you decide to be satisfied with an average efficiency of 45% or to try for 90% will depend mostly on how much money you would like to invest. Either way you'll be saving money in the long run.

How Much Does a Solar Water System Cost?

Let's discuss collectors first, since they might represent the largest expense. How much you will have to spend depends on a number of things: where you live, how many people there are in the family, how efficient you want your system to be, and whether you plan to buy pre-fabricated collectors or build them yourself. (Chapter 5 gives you details on how to build your own collectors.)

A typical "flat plate" collector — the kind used for solar water heating — is about 3 feet wide and 8 feet long, though this can vary from manufacturer to manufacturer. (You might also decide — if you're making your own — that these dimensions don't fit your situation.)

In any case, the typical surface area of a collector is 24 square feet. (Not *all* collectors have 24

square feet, remember.) If your calculations tell you that you're going to need two collectors, it's easy enough to figure out that you'll have 48 square feet of collector surface.

The cost of a ready-made solar collector is $10 to $12 a square foot. Averaging this to $11 a square foot your cost for 2 collectors would be $528 (11 x 48). If you build your own collectors, on the other hand, you should be able to do it for $4 to $4.50 a square foot — even if you buy the most critical and hard-to-make component, the collector plate itself. (More on the collector plate in Chapter 3.) At $4.50 a square foot, the cost of 2 collectors would be only $216. So it is worthwhile — and not that difficult — to build your own.

What about other costs? There are all kinds of variables here too, but we can make a few general assumptions. Let's say you're going to build the most sophisticated system recommended in this book, but let's also say that you're going to buy parts — other than the collectors — from a plumbing supply house instead of buying your whole system in a pre-fabricated "package" from a company that specializes in solar equipment.

First, you'll need a storage tank — probably 120 gallons — with a heat exchanger in it. The plumbing supplier or distributor should charge you about $290 for this. Next you'll need a circulation pump — another $65; next, some thermostatic controls. This sounds expensive, but it's not bad — about $40. You'll also need some plumbing — tubing, fittings, assorted valves and perhaps an expansion tank — along with some special anti-freeze. These shouldn't run more than $100.

All told, these items should cost under $500 — $495 to be exact. (Remember now, these prices are for 1978.) In other words, if you do most of the work yourself, and buy directly, eliminating the middle-men who would be involved if you bought a "prefab" system, you should be able to put together an excellent solar water heater for as little as $750, allowing a few extra dollars for things like roofing cement, caulking compound, nails, screws, flashing and so forth ($495 plus $216 = $711).

Will you wind up with an inferior system if you build it yourself? You shouldn't. The components will be the same or better than those in a prefabricated package; you are just buying them at a cheaper rate. Will a pre-fabricated package last longer? Not if you design your own carefully, use the best materials you can get for the money, and treat the system well. In fact, if you build it yourself, you should know your system intimately, and should be able to do most of the maintenance work yourself.

If you don't enjoy shopping around, and want to avoid the hassle of doing the work yourself, tack about 45% on to the plumbing costs listed above. This means that your system will cost somewhere between $1,200 and $2,000. Beyond that, you should add another $500 to $700 for installation. In short, your pre-fabricated system is going to cost $2,500 or more. To put it another way, having a contractor do the footwork and building for you will cost you about 3 times as much as doing it yourself.

How Long Does it Take to Pay Off a Solar Water Heater?

We have already figured out that it can cost a family of 5 about $348.43 a year to heat water to 140 degrees Fahrenheit with electricity. If this same family could build a solar water heating system for $750 and that system were 75% efficient, they would be saving $261.32 a year in electrical utility costs ($348.43 × .75 = $261.32). Dividing this figure into $750, we can compute that it will take 2.8 years to pay for the system (with inflation it may take even less time).

If the family spent $2,500 to have the system installed by someone else, it would take 9.5 years to pay it off. Let's keep in mind though, that electricity is the most expensive way to heat water. Fuel oil and natural gas are somewhat cheaper. (It would take 8.2 years to pay off a homemade system that was previously run entirely with oil, and 11.2 years to pay off the same system if it were fired by natural gas.)

To summarize all this in less mathematical terms, an average family, with a system of average efficiency, barring any major breakdowns

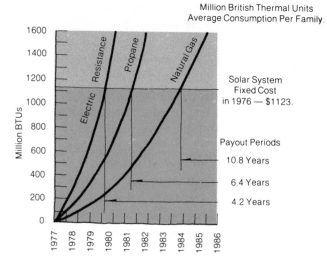

Graph 1. This graph, calculated by Paul Shippee of Colorado Sunworks, P.O. Box 455, Boulder, projects the costs of heating water with electricity, propane, and natural gas over the next several years. It also gives you an indication of how long it should take to pay for a solar water heater. Shippee's figures, although they vary slightly, are very close to our own.

and additional expenses, should be able to pay for its solar hot water heating system in 5 to 10 years — which is what you really wanted to know in the beginning.

Are There Other Considerations?

Yes. When you come right down to it, there are two big reasons for putting in a solar water heating system: (1) It will save you money, and (2) the fuel you save will benefit society as a whole. Unfortunately our society — formally at least — still doesn't seem to recognize this second benefit. The whole solar energy field screams for more special loans and grants for down-to-earth research and development, and for tax incentives which would encourage consumers like you and me to put solar systems in our homes. But these are very slow in coming.

Incredible as it may seem, there are many places in the United States where installing a solar system actually raises a family's real estate

10

taxes. In some cases, this increased tax offsets whatever they might be able to save in fuel costs. What this amounts to is that government is penalizing people for taking the bull by the horns and doing what they can to conserve the nation's energy. Wherever this happens, there's obviously something wrong with the system. Consult your local lister or tax assessor to find out where you will stand if you decide to change over to solar water heating.

According to Arthur K. Little Associates, solar energy research specialists, manufacturing solar heating equipment is going to be big business by 1985. That's terrific, but unfortunately right now there is still a need for more performance and maintenance standards in the industry. Some products simply have not been tested enough, and the manufacturers make overly optimistic claims about them.

This is only part of the reason why fairly standard, tried-and-true plumbing and off-the-shelf collector components which you can understand and install yourself are the best investment. Semi-honest, fast-buck contractors are descending on the solar market like vultures. There have already been several lawsuits brought against such operators in Connecticut and elsewhere. Others, whose intentions are fine, still have a lot to learn about the field, and are likely to make mistakes as they cut corners and try to increase their profits.

There have been horror stories in *The New York Times* about how solar hot water systems break down because of air locks in the water pipes, improper fittings, putting circulation pumps in backwards, faulty wiring, setting timers so that units worked only at night when there was no sun, and about leaks in pipes that froze because of poor insulation. Most of these could have been avoided had the installers had just a little more background and experience.

A solar water system should have a minimum life expectancy of 20 years or more. If it's put together carefully with the highest quality materials, if precautions are taken against corrosion and freezing, and if it's looked after as regularly as you would your car, there's no reason it shouldn't last a lifetime. After all, it has few moving parts, and anything that might wear out can be readily replaced.

How can you afford the expense of converting to solar hot water? Home improvement loans for solar systems have been difficult to come by, not only because the field seems new and people are skeptical, but because there have been no accepted standards set up by agencies like the Federal Housing Authority and the Veteran's Administration. But as time goes on, and the energy pinch gets tighter, bankers should become more and more sympathetic to loan requests for solar installations.

There *are* some proposals for federal tax credits in the works now. If passed by Congress, these would assume up to 40% of the cost of a new solar system. But this legislation is still in the planning stages. Check later to see what results. In some states, as of June 1977, you could apply for a $400 federal grant toward installing a solar water heating system. Ten thousand grants will be awarded in all. The ten states that offer this program are:

Connecticut	New Jersey
Florida	Pennsylvania
Maryland	Rhode Island
Massachusetts	Vermont
New Hampshire	and parts of New York

Check with your state energy office for more details.

More federal and state money should be available soon as our demand for solar energy increases. To stay abreast of what is available for funds, or to keep up to date on what is happening in the solar energy field, write or call the Solar Heating and Cooling Institute, Box 1607, Rockville, Maryland, 20850. They even have a toll-free number — 800-523-2929. If you call, they'll bombard you with helpful information and data.

CHAPTER 2

Some Words About the Sun

Ancient and "primitive" peoples once worshipped the sun as the ultimate giver-of-life. Some philosophers and serious futurists around today would say that we're coming back to that. Somewhere along the way we just got misled.

For some reason that no one will ever explain fully, our universe is dotted with immense balls of energy which we have chosen to call "stars." If we consider the universe to be infinite, then it must follow that there is an infinite amount of energy out there. Our particular star — about which we circle every 365.25 days — is expected to keep burning for at least another 4 billion years. Al-

though our sun may be finite as an energy source, its end is nowhere in our sight.

The sun is already 8 to 10 billion years old, as near as we can guess, and it has been staring us in the face ever since the beginning of Man's life. (It will probably still be there long after we've gone and something else has taken over this planet we think we own.)

It's ironic that since the start the sun has been saying, "Here I am!" But our minds got distracted or somehow thrown out of whack because we've kept saying, "Don't bother me, Sun, I'm looking for energy."

Once in a while someone would pause briefly as if to think, "Sun, you're a ball of energy today," and even go so far as to dream up ways to harness some of the sun's radiation. Others of us have looked on with amused interest, but finally said, "That's cool, man, but coal (wood, oil, gas, nuclear) is cheaper and easier to get."

Figure 3. The sun gives off rays of different wave lengths. Infrared rays, the ones that give us heat, have longer wave lengths. We are pelted less often with solar energy units — called "photons" — when the waves of solar energy are longer. Ultraviolet rays have shorter wave lengths, which means that more photons strike the earth at a more rapid rate.

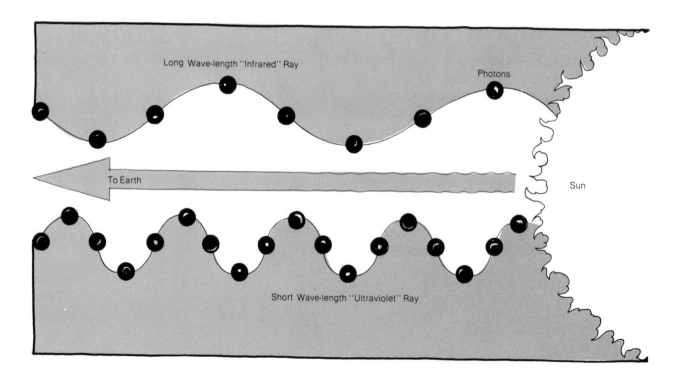

The sun's energy is all around us ... everywhere. We just can't always see it. The sun moves through space — bringing us along on a gravitational leash — giving off power in all directions. We receive about 1/200 millionth of its total output at any given time. The energy originates in a process that we understand as "thermonuclear fusion," which is like the reaction that takes place in a hydrogen bomb. Solar scientists estimate that the temperature near the sun's core is 45 million degrees Fahrenheit, although it's several million degrees cooler than that on the solar "surface."

Energy is "beamed" to us over a distance of roughly 93 million miles. The trip, at 186,000 miles per second, takes just over 8 minutes. Each wavy beam of sunlight consists of tiny solar energy units called "photons," and these photons are being fired at us as though out of a super-high-speed machine gun. When they hit us more frequently — because they come at us in shorter wave lengths — their energy is greater than when they strike us less often — having arrived in longer wave lengths. That's just another way of saying that shorter wave lengths mean more photon "bullets" and thus more power (Figure 3).

Plants catch photons, but they're only as efficient about it as they need to be — using about 1 percent of what reaches them. Through the wonder of photosynthesis they change these photons into chemical energy and then store it (Figure 4). We eat plants — or other animals that eat plant life — and also burn certain wood plants to get at some of the energy they have saved up. Yet the vast majority of plants die a natural death (without being eaten), then decay and are compressed by the weight of the generations that follow. After millions of years they become fossil fuels — different forms of stored-up chemical energy from vegetation (Figure 5).

The sun not only sets off the processes that create food and fuel, it also directly influences other "kinetic" energy sources, such as the winds, the tides, rain and thermal forces that we've only just begun to wonder how to use.

The sun is at the heart of it all, but in the twentieth century our civilization has become so complex and we've become so disoriented that we lose sight of this fact. It's as though we are

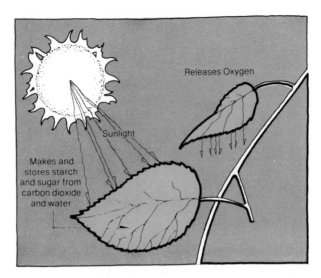

Figure 4. Photosynthesis is the chemical miracle that goes on all around us every day. Green plants, which contain chlorophyll, convert carbon dioxide and water into starches and sugars with the help of sunlight. Oxygen is a by-product. In a sense, the plant becomes a storage area for solar energy, and there are times when this energy is stored for millions of years. It gets released only when we burn fossil fuels.

living off a dwindling savings account — chewing our nails at how fast it's going — while we ignore the steady income of energy that pours down on us every day. It's as though in searching out the remaining supplies of fossil fuels (because that's all we think we know how to do) we're wandering in the woods looking for this big stash of change that somebody has missed, while there are tons of pennies raining all around us.

It's time to turn back to the sun.

How Much Energy Do We Get From The Sun?

The diameter of the sun is much larger than that of the earth — by about 110 times, in fact. One sun "g" would be equivalent to 28 earth "g's," which means we would weigh 28 times as much as we do now if we could somehow exist on the sun. Huge as it is, the sun is converting mass into energy at a rate of some 4.7 million tons per second and it doesn't seem to be getting much smaller (Figure 6).

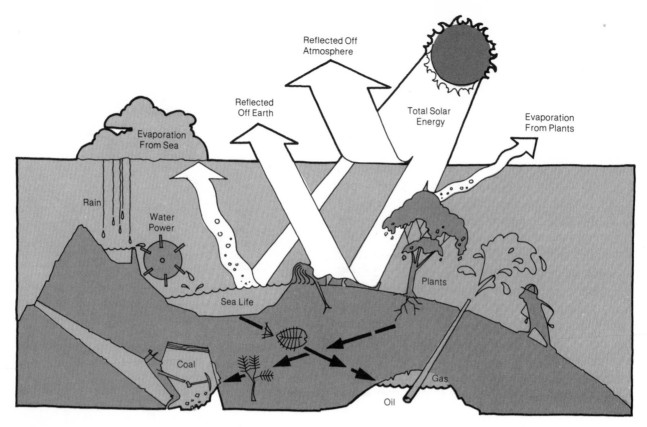

Figure 5. All of the ecosystems on earth are dependent on the sun. In comparison to our supplies of fossil fuels, it is by far our largest natural resource. Solar energy is stored in fossil fuels, but that's not a very efficient means of storage because only a miniscule percentage of plant or animal life actually becomes coal, oil, or natural gas.

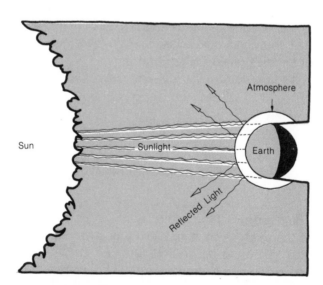

Figure 6. The earth receives only about 1/200 millionth of the total energy of the sun. Even then, a large portion of solar radiation is reflected back into space by our atmosphere. But what does get through to us is more than enough to supply all of our energy needs.

Our "furnace in the sky" shoots power into our atmosphere at a fairly consistent rate. This regular and dependable energy supply is called the *solar constant,* and it can be measured. Just one square foot of the outside of the earth's protective air layer gets a whopping average of 430 BTU's per hour (Figure 7).

Energy expert Wilson Clark estimates that if all of the solar constant got through the atmosphere, and if we could collect it at 100% efficiency, an area the size of the United States would absorb enough energy in just 32 minutes to supply the world's complete energy needs for a year.

That the atmosphere absorbs a large percentage of solar radiation is good news as well as bad. The shortest waves of solar radiation, *ultraviolet rays,* have the greatest energy content. Most of these are screened out by our all-important ozone layer, making life on earth possible. If these rays were to reach us directly and in

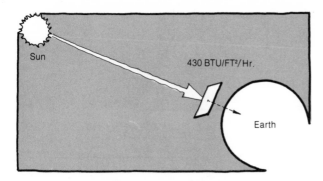

Figure 7. The *solar constant* is the amount of solar energy that hits the outside of the earth's atmosphere. One square foot of surface that is perpendicular to the sun's rays in the outer atmosphere receives 430 BTU's per hour. This is usually written as 430 Btu/ft²/hr. A figure like this can be misleading because not all of this energy makes it through to the earth's surface.

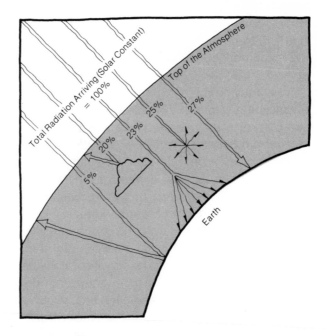

Figure 8. Here's what happens to the solar constant once solar rays hit our atmosphere: (A) 5% is reflected off the earth itself. (B) 20% is reflected off the clouds, (C) 23% reaches the ground as diffuse sunlight, (D) 25% is absorbed in the atmosphere, and (E) only 27% actually reaches us as direct parallel rays.

full force, we would all be sitting in a gigantic microwave oven that would cook us — and most living things around us — in short order.

A lot of solar energy is reflected off, or absorbed by, our atmosphere and bounced back into outer space by clouds. But those solar rays that hit the earth on a very clear day are, for the most part, parallel to each other. When there is haze, cloud cover, smog or dust in the air, the parallel pattern is broken and the rays are deflected off in many different directions by these particles of water or dust in the atmosphere. This is why heat and light often seem to come at us from all parts of the sky. "Diffuse" radiation, as this is called, is still very strong and is useful to us if we have the right kind of solar collector (Figure 8).

After most of the strongest and most harmful rays have been filtered out, and a large share of the sun's energy has been dissipated or turned away, the average solar intensity on the ground is still 1,400 BTU's per square foot per day. That's enough energy in a square mile to equal the productivity of a large hydro-electric power plant (Table 4).

There have been all kinds of mind-boggling statistics to describe the amount of solar energy that falls on us. D.S. Halacy, Jr., in *The Coming of Age of Solar Energy,* says that man consumed 90 trillion horse-power hours of energy in 1972, but that in that same year 1.5 *million* trillion

TABLE 4

AVERAGE SOLAR ENERGY

	Megawatts of Electricity per Square Mile per Day		
	December	March	June
New York and Chicago	130	330	575
Southern California and Arizona	260	420	730
Florida	260	420	575
Nevada	180	420	675

After Wilson Clark, *Energy for Survival*

Table 4. This simple table shows the average amount of solar energy that falls in a square-mile area in selected parts of the United States. In this case, the energy is expressed as megawatts — 1,000 kilowatts. A typical modern electrical power plant only generates a few hundred megawatts a day. Some of the largest manage to put out 1,000 per day.

horse-power hours of solar energy fell on Mother Earth. In electrical terms that would mean that for every square foot of earth, there would have to be at least 1,000 watts generated each day.

What may be more important for the purposes of collecting solar energy to heat water is that the sun comes to us in three different forms: (1) *direct parallel rays;* (2) *diffuse radiation,* which comes to us from all directions after being glanced off clouds, dust or moisture particles in the air; and (3) *reflected radiation,* which has been bounced straight off some surface.

Some ultraviolet rays do get through to us. We can't see these rays, but our skin can feel them after a couple of hours on a sunny beach. *Visible sunlight,* what we think of as brightness during the day, has much longer wave lengths, relatively speaking, than ultraviolet light. If we look through sunlight that has been bent into a rainbow by a prismatic effect in the air, we see that visible light consists of blue, green, yellow and red hues. *Infrared* rays, which give us heat, have longer wave lengths still, and like ultraviolet rays they are invisible to us.

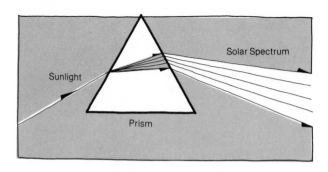

Figure 9A. When light rays are passed through a prism and "refracted," we can see the colors red, orange, yellow, green, blue indigo and violet. This is visible sunlight.

The sun's energy that actually reaches us is about 80% visible light, 4% ultraviolet light, and about 16% infrared (Figure 9, 9A). A very high percentage of this great energy can be captured by you and me if we have the right equipment.

How Does Sunlight Hit the Earth?

One of the things you will have to consider as you make decisions about your own involvement with solar energy, is the latitude of your home. You probably don't need to be told that the farther you go from the equator — that imaginary line that slices the globe into a top half and a bottom half — the weaker the sun's energy becomes.

What many people *don't* realize is that the more horizontal lines — "parallels" of latitude — there are on the map between you and the equator, the farther sunlight has to travel to reach you, and the more energy is lost. The reason we can look at the sun at sunrise and sunset is that we're looking at it over the curvature of the earth. Much of the light is being absorbed by the atmosphere. At noon, when the sun is more overhead, its rays have to pass through fewer fathoms of air before they reach our eyes (Figure 10).

But it's not quite so simple as that. There are great differences in *insolation* from place to place along the same latitude. ("Insolation" is to sunlight what the measurement called "rainfall"

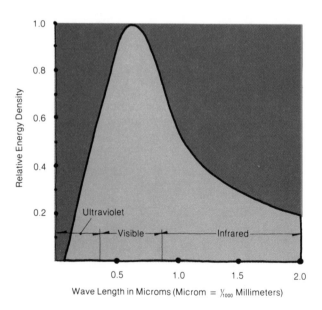

Figure 9. This strange-looking curve represents the "solar spectrum." Nearly everything that happens on earth is dependent on this narrow range of solar radiation. *Visible sunlight,* as you can see, is most common, and *ultraviolet rays* are screened out, for the most part, before they reach the ground. We can't see either ultraviolet rays or *infrared light.* (After Wilson Clark, *Energy for Survival)*

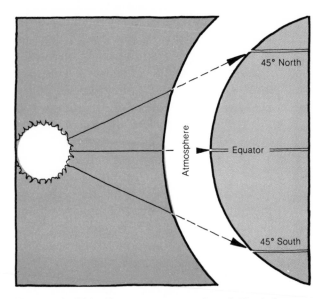

45° North

Atmosphere

Equator

45° South

Figure 10. This diagram suggests how latitude has an effect on insolation. As we go further and further from the equator in either direction, sunlight has to pass through more and more of the atmosphere to reach us. By the time it reaches 45 degrees north or south, much of its force has been stolen by the atmosphere.

is to rain.) These differences are the result of altitude, local climatic conditions, and seasonal changes (Table 5).

Places with little rainfall tend to be good places to collect an abundance of solar energy — because there are few clouds. But dusty and windy places are bad because the particles in the air block a lot of rays. Locations near the sea are apt to have lower insolation than places at higher altitudes where clouds, haze and foggy conditions are less likely to rob sunshine. (Places in New England may have as many as 10 cloud-covered days in a row, giving them a year's total of 40% diffuse sunlight on the average. But the mile-high city of Denver, Colorado, even in

Table 5. This is a list of United States Weather Service Stations currently recording sunshine. As you can see, they are still few and far between, but there will be more as more of us recognize the importance of solar energy.

TABLE 5

Alaska	*Idaho*	*Nevada*	*South Dakota*
Annette	Boise	Ely	Rapid City
Barrow	*Illinois*	Las Vegas	*Tennessee*
Bethel	Argonne	Reno	Nashville
Fairbanks	*Indiana*	*New Mexico*	Oak Ridge
Matanuska	Indianapolis	Albuquerque	*Texas*
Arizona	*Iowa*	*New York*	Brownsville
Page	Ames	Astoria	El Paso
Phoenix	*Kansas*	Geneva	Fort Worth
Tucson	Dodge City	Ithaca	Midland
Arkansas	Manhattan	New York Central Park	San Antonio
Little Rock	*Louisiana*	*North Carolina*	*Vermont*
California	Lake Charles	Cape Hatteras	Burlington
Davis	*Maine*	Greensboro	*Washington*
El Centro	Caribou	*North Dakota*	Seattle
Fresno	*Michigan*	Bismarck	Tacoma
Inyohem	East Lansing	*Ohio*	*Wisconsin*
Los Angeles	Sault Ste. Marie	Cleveland	Madison
Riverside	*Mississippi*	*Oklahoma*	*Wyoming*
Santa Maria	Columbia	Oklahoma City	Lander
Colorado	*Montana*	*Oregon*	Laramie
Grand Junction	Glasgow	Medford	*"Proposed" Stations*
District of Columbia	Great Falls	Portland	Boulder, Colorado
Washington	*Nebraska*	*Pennsylvania*	Columbia, Missouri
Florida	North Omaha	State College	Salt Lake City, Utah
Apalachicola		*Rhode Island*	Sterling, Virginia
Miami		Newport	
Tallahassee			

January — the worst month for sun — will have 67% *direct* radiation. See Table 6).

The best area to collect sunlight is between 15 and 35 degrees either north or south of the equator, because most of the radiation is quite direct and there are few clouds. (It's no accident that this is where the world's great deserts are.) Next best is 15 degrees immediately below and above the equator, except that here there are plenty of clouds and lots of humidity in the lower atmosphere. Once you get beyond 35 degrees north or south, there's a strong effect on insolation caused by changes in the seasons. This may be the main reason why 75% of the world's population lives below 40 degrees north latitude and above 40 degrees south latitude. Beyond these points winter really takes its toll (Figure 11).

Remember, the sun's rays are hitting a curved surface — the side of our round earth. So the angle at which the sun's rays hit the earth gets lower the further we go from the equator. (If any receiving surface is tilted 45 degrees away from directly headed rays of sunlight, the intensity of radiation will be only 71%. If the surface is tipped

Table 6. January is the month when there's likely to be the least amount of sunshine. Notice that Portland, Oregon is sunny only 19% of the time in January, while Denver has direct radiation 67% of the time.

TABLE 6

PERCENT OF POSSIBLE SUNSHINE IN JANUARY (SELECTED MAJOR CITIES)

State and City	%	State and City	%	State and City	%	State and City	%
Alabama		Pueblo	71	Indiana		Springfield	—
Anniston	50	Connecticut		Evansville	38	Worcester	—
Birmingham	45	Bridgeport	—	Fort Wayne	35	Michigan	
Mobile	51	Hartford	46	Indianapolis	36	Alpena	28
Montgomery	51	New Haven	52	Terre Haute	44	Detroit	31
Arizona		Delaware		Iowa		Escanaba	41
Flagstaff	—	Wilmington	—	Davenport	—	Flint	—
Phoenix	76	District of Columbia		Des Moines	54	Grand Rapids	26
Tucson	78	Washington	46	Dubuque	48	Lansing	—
Winslow	—	Florida		Keokuk	—	Marquette	28
Yuma	84	Apalachicola	58	Sioux City	56	Saginaw	—
Arkansas		Jacksonville	57	Kansas		Sault Ste. Marie	32
Fort Smith	42	Key West	68	Concordia	60	Minnesota	
Little Rock	42	Miami	66	Dodge City	61	Duluth	46
California		Pensacola	55	Topeka	51	Minneapolis	49
Bakersfield	—	Tampa	60	Wichita	61	St. Paul	—
El Centro	—	Georgia		Kentucky		Mississippi	
Eureka	39	Atlanta	48	Louisville	41	Meridian	—
Fresno	49	Augusta	56	Louisiana		Vicksburg	45
Long Beach	—	Brunswick	—	New Orleans	49	Missouri	
Los Angeles	71	Columbus	—	Shreveport	47	Columbia	52
Needles	—	Macon	54	Maine		Kansas City	51
Oakland	—	Savannah	54	Augusta	—	St. Louis	46
Pasadena	—	Idaho		Bangor	—	Montana	
Sacramento	39	Boise	39	Eastport	45	Billings	47
San Bernardino	—	Lewiston	—	Portland	55	Butte	—
San Diego	67	Pocatello	35	Maryland		Havre	49
San Francisco	53	Illinois		Baltimore	48	Helena	51
San Jose	54	Cairo	42	Cumberland	—	Kalispell	—
Colorado		Chicago	42	Massachusetts		Missoula	29
Denver	67	Moline	42	Boston	48	Nebraska	
Durango	—	Peoria	47	Fitchburg	—	Lincoln	57
Grand Junction	53	Springfield	38	Nantucket	43	North Platte	62

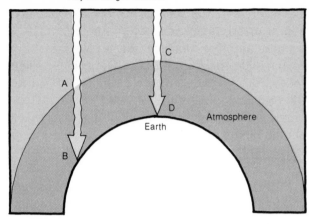

Ray AB Longer Than CD

C
A
D Atmosphere
Earth
B

Figure 11. For all practical purposes the sun's rays are parallel — until they get deflected by particles in the atmosphere. Ray AB in this diagram is weaker by the time it hits the ground than ray CD. This is not only

60% away from the source, the radiation drops way down to 50% (Figure 12).

This is why as we move further north (or south) we have to plan carefully how we will orient a solar collector in relation to the sun. We want rays to hit the collector at close to right angles as much of the time as possible. At latitudes well above 35 degrees, a flat plate collector that's angled properly should have no trouble heating water to 140 degrees for at least a few hours a day, even in winter. As latitudes get higher, solar

because AB has to pass through more air, but because it hits the side of the earth at a flat angle instead of straight on. CD, on the other hand, hits the ground at right angles, so its effect is felt more strongly.

State and City	%	State and City	%	State and City	%	State and City	%
Omaha	54	North Dakota		Block Island	45	Salt Lake City	46
Valentine	62	Bismarck	54	Pawtucket	—	Vermont	
Nevada		Devils Lake	52	Providence	49	Burlington	34
Reno	64	Fargo	48	South Carolina		Rutland	—
Winnemucca	53	Williston	51	Charleston	58	Virginia	
New Hampshire		Ohio		Columbia	53	Cape Henry	40
Concord	48	Akron	—	Greenville	53	Lynchburg	48
Manchester	—	Cincinnati	40	South Dakota		Norfolk	50
Portsmouth	—	Cleveland	28	Huron	52	Richmond	49
New Jersey		Columbus	36	Rapid City	55	Roanoke	—
Atlantic City	51	Dayton	38	Sioux Falls	—	Washington	
Jersey City	—	Sandusky	34	Tennessee		North Head	28
Newark	—	Toledo	32	Chattanooga	43	Seattle	28
Trenton	48	Youngstown	—	Knoxville	42	Spokane	26
New Mexico		Oklahoma		Memphis	45	Tacoma	21
Albuquerque	69	Oklahoma City	57	Nashville	42	Walla Walla	24
Roswell	—	Tulsa	49	Texas		Wenatchee	—
Santa Fe	—	Oregon		Abilene	59	Yakima	34
New York		Baker	41	Amarillo	65	West Virginia	
Albany	42	Eugene	—	Austin	48	Bluefield	—
Binghamton	32	Medford	—	Brownsville	48	Charleston	—
Buffalo	32	Portland	19	Corpus Christi	49	Huntington	—
Canton	—	Roseburg	25	Dallas	46	Parkersburg	29
New York	51	Pennsylvania		Del Rio	53	Wheeling	—
Oswego	19	Altoona	—	El Paso	75	Wisconsin	
Rochester	32	Erie	22	Fort Worth	54	Green Bay	46
Syracuse	31	Harrisburg	44	Galveston	50	La Crosse	—
North Carolina		Oil City	—	Houston	45	Madison	46
Asheville	48	Philadelphia	45	Palestine	46	Milwaukee	38
Charlotte	52	Pittsburgh	32	Port Arthur	47	Wyoming	
Greensboro	49	Reading	44	San Antonio	46	Cheyenne	61
Raleigh	49	Scranton	37	Utah		Lander	63
Wilmington	57	Rhode Island		Modena	—		

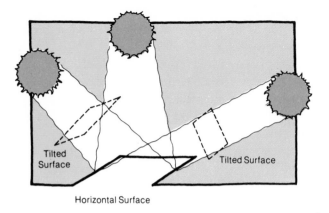

Figure 12. There is a tremendous difference between the energy that is received by a surface that is square to the sun's rays and one that is at an angle. This is why solar collectors that are far from the equator must be tilted to absorb the maximum amount of radiation for as many hours as possible during the day.

collectors have to be tilted more and more toward vertical, because at that point a horizontal collector will absorb next to nothing in the way of heat.

How Do the Seasons Affect Insolation?

Years ago our parents or some elementary school teacher taught us that the earth goes around and around and that the sun shines on us from 93 million miles away. This is why we have day and night. What they may or may not have explained was that the earth moves around the sun at an average speed of about 18.5 miles a second, and that on December 21 we are about 89.8 million miles from the sun, but 95.9 million miles from the sun on June 21. (This is because the sun isn't always in the exact center of our orbit.) Our childish logic might have told us that it should be the other way around: if we were closer in December, it should be warmer then — summer!

What we also may not have understood as very young children, but could grasp later on, was the fact that the earth's axis — about which we go "round and round" — is tilted 23.5 degrees from vertical. (What we may recall most vividly from those early astronomy lessons is that the earth's axis points directly at the north star.) This axial tilt — not our distance from the sun — is what causes changes in the seasons (Figure 13).

As we spin once a day, 365.25 times, we move in a path around the sun one time. (One revolution accounts for a year of what we call "time," and the fact that each year has an extra quarter of a day explains why we need a leap year once every 4 years.) For some reason, as we get closer to the sun, our speed slows down slightly and then quickens again as we get further away. So

Figure 13. The earth spins on a tipped axis as it moves around the sun. This is why we have changes in seasons. On June 21, although we are actually further from the sun, we are tilted toward it — that's if we live in the northern hemisphere. On December 21, we are tipped away from the sun, so its rays don't hit us directly.

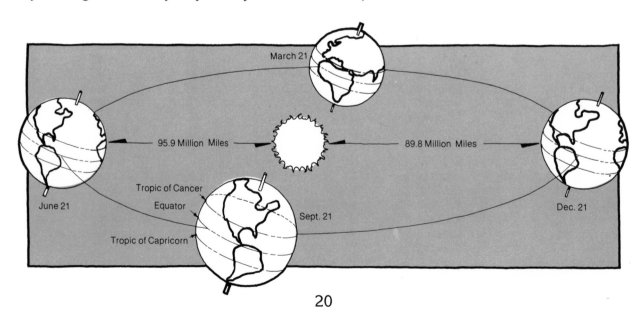

we're not always orbiting the sun at the same precise rate of speed.

On June 21 — the *summer solstice* — we in the northern half of the world are tilted most toward the sun so it hits us most directly. This is the day when we receive sunlight for the longest amount of time. On December 21, we are tilted farthest away from the sun. Then we get the fewest hours of sunshine and those rays we do get hit us at a pretty flat angle.

On March 21 and September 21 — the spring and fall *equinoxes* — there are equal amounts of sunlight and darkness. The Tropic of Cancer (23.5 degrees north), that weird parallel on the globe that seems out of spacing with the other latitude lines, marks the point where the sun is directly overhead at noon on June 21. The Tropic of Capricorn (23.5 degrees south) describes the same path that most direct sunlight takes in the southern hemisphere on December 21. When it is summer in the southern half of the world, of course, it is winter in the north (Figure 14).

That's how things look, theoretically at least, from outer space. But here on earth we're looking at it from the opposite end, which gives us a very different perspective. What we earth beings see is this: On June 21 the sun appears higher in the sky at noon than it does at any other time of the year. It *is* higher, in relation to the horizon, and this angle between the horizon and the sun is expressed as the sun's *altitude* (Figure 15).

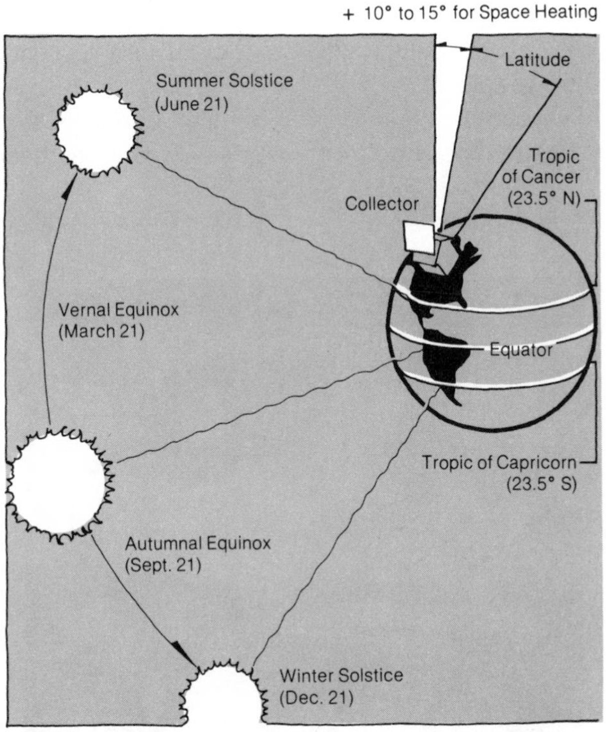

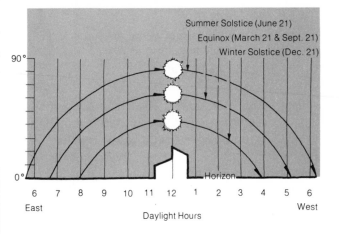

Figure 15. We see more of the sun in June than we do in December. In the summer it is not only higher in the sky; it is there for more hours.

Figure 15A. The sun's position relative to the horizon is called its *altitude*. The sun's altitude is expressed in degrees, not in distance.

Figure 14. The Tropic of Cancer indicates the northernmost point on earth where the sun can be directly overhead. This happens on June 21. On December 21 it is directly above the Tropic of Capricorn. This diagram shows the seasonal relationships between the sun and the earth. The normally recommended angle for a fixed solar collector is your latitude plus 10 to 15 degrees.

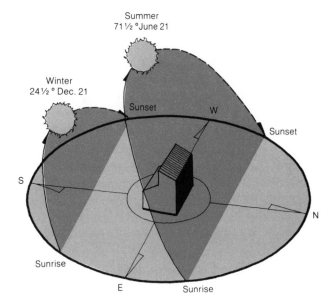

21

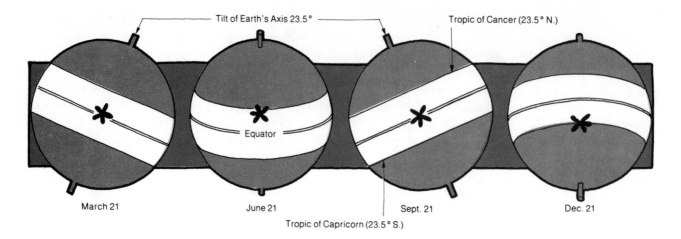

Tilt of Earth's Axis 23.5°   Tropic of Cancer (23.5° N.)

Equator

March 21   June 21   Sept. 21   Dec. 21

Tropic of Capricorn (23.5° S.)

Figure 16. X marks the spot where the sun is directly overhead at four different times of year.

The sun is at its lowest altitude on December 21 — the *winter solstice.* On March and September 21 its position in the sky is midway between the other two extremes. We not only *see* less of the sun in December because it is around for fewer hours, its intensity is diminished because of its low altitude. This is why we have winter (Figure 16).

Don't be confused by insolation maps like the one in Figure 17. Here solar energy is being measured and described in "Langleys" rather than BTU's. The Weather Service is simply using the language of the metric system instead of the old English system. A *Langley* is the amount of heat it takes to heat one gram of water one degree centigrade. To convert Langleys per day to BTU's just multiply the number of Langleys by 3.69 (Table 7).

One more thing to keep in mind: Langleys are measured on a horizontal surface at any weather

How Does All of this Relate to Solar Water Heating?

Before you make a commitment to a solar heating system of any kind you should study how much sunlight falls on your area at all times of the year. Solar insolation maps that are even more accurate than U.S. Weather Service maps can be found in a book called *World Distribution of Solar Radiation,* by Smith, Duffie and Löf. For a copy of this title write to the University of Wisconsin, Engineering Experiment Station, 1500 Johnson Drive, Madison, Wisconsin 53706, and send them a check for $6.00. You may also get more specific climatic information from your nearest Weather Service office. And *The Old Farmer's Almanac,* although not always precise, is another good source of general weather data (Figure 17).

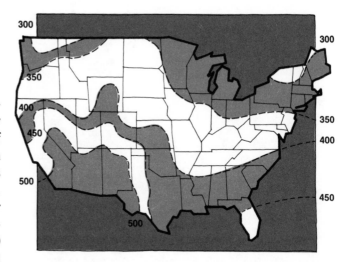

Figure 17. This map, similar to one from the U.S. Weather Service, shows the average annual insolation in the continental United States. The figures are in Langleys per day, not in BTU's per square foot (1 Langley = 3.69 Btu/ft²). Other maps available from the Weather Service give a monthly breakdown of insolation.

station, meaning that a solar collector angled directly toward the sun could be expected to pick up a great deal more energy at most times of the year than the insolation map suggests (Graph 2).

Before you decide whether or not a solar system is feasible, learn all you can about your local weather idiosyncracies — specifically how many cloud-free days there are each month or how smoggy it gets at various times of the year. (Smog reduces insolation by as much as 15 to 20%.)

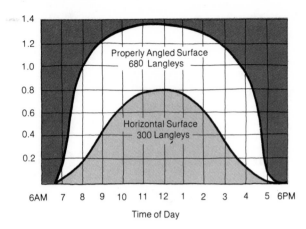

Graph 2. This revealing graph, borrowed from Farrington Daniels's *Direct Use of the Sun's Energy*, records the amount of radiation received on a normal surface and a horizonal surface in late October in Madison, Wisconsin, Latitude 43 degrees N. Note how the tipped surface caught more than twice the number of Langleys.

TABLE 7

CONVERSION FROM LANGLEYS PER DAY TO BTU'S PER SQUARE FOOT PER DAY

Langleys/day	Btu's/sq. ft./day
100	369
150	554
200	738
250	922
300	1,110
350	1,290
400	1,475
450	1,660
500	1,845
550	2,130
600	2,220
650	2,400
700	2,580
750	2,760
800	2,950
850	3,140

Let the nearest Weather Service Office help you, rather than try to collect weather data yourself. Weather monitoring equipment is expensive, and your reading over the course of a year or two will probably not be as accurate as the long-range averages the U.S. Weather Service has gathered.

Once you become a sun worshipper (or at least an admirer) to the extent that you realize how much it's giving you — probably enough to heat most of your hot water where you are — you can begin planning how to collect it.

heliochemical conversion, because solar energy is being changed over to stored chemical energy. Gathering energy through a man-made solar collector involves a *heliothermal* change, because radiation is absorbed by a black surface and turned into heat. This process is very simple, very effective, and it is what Chapter 3 is all about.

CHAPTER 3

Harvesting Energy from the Sun

What Are Focusing Solar Collectors?

The Greek word for sun is *Helios.* As we all know, much of our English language has evolved from the ancient Greek, and many English words have Greek parts. So any word that has the prefix *helio-* probably has something to do with the sun.

A *helioelectrical* process is a way of converting sunlight to electricity — something we're just learning to do efficiently. Photosynthesis, mentioned in the last chapter, is an example of a

Focusing collectors — or *solar concentrators* as they're sometimes called — redirect the sun's parallel rays to a specific focal point or "target."

Figure 18. A focusing — or concentrating — collector changes the direction of parallel solar rays so they zero in on a given focal point. Double convex lenses, such as those in magnifying glasses, do this directly. A parabolic mirror, on the other hand, reflects concentrated rays back to a target in front of the reflector. Either way sunlight is intensified tremendously.

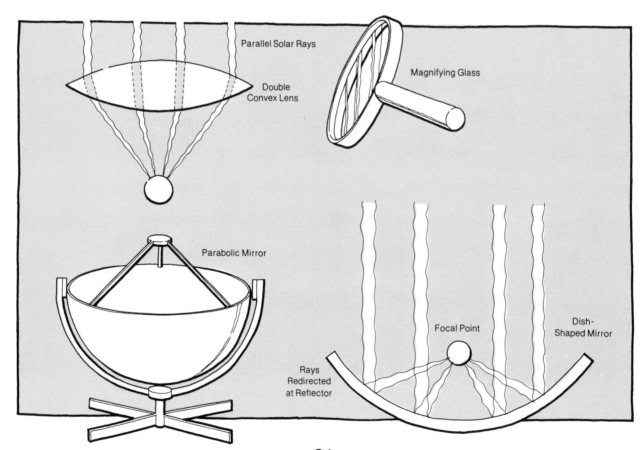

Parallel Solar Rays

Double Convex Lens

Magnifying Glass

Parabolic Mirror

Rays Redirected at Reflector

Focal Point

Dish-Shaped Mirror

24

Anybody who has ever lit a fire with a magnifying glass has used a solar concentrator. A magnifying glass is actually a double convex lens, which bends parallel rays and makes them converge on a central point. The bead of light thrown by the glass can be made hot enough — by adjusting the height of the glass — to ignite paper or even light kindling (Figure 18).

Another common type of focusing collector is a *parabolic mirror,* a bowl-shaped reflector that directs solar rays to a central target right in front of the mirror itself. A good silvered-glass mirror or polished metal surface shaped like the inside of a car's sealed beam headlight should reflect about 92% of the light that hits it and shoot it all to one spot. The advantage of any focusing collector is that it can generate tremendous temperature — as high as several thousand degrees — because it intensifies radiation hundreds of times (Figure 18A).

But there are serious *dis*advantages to focusing collectors. First, they must point exactly toward the sun or they won't work at all. This means that they either have to be moved by hand or be hooked up to a device called a *heliostatic mount* that tracks the sun all day long. Second, most solar concentrators can't focus diffuse light, making them almost worthless when there is

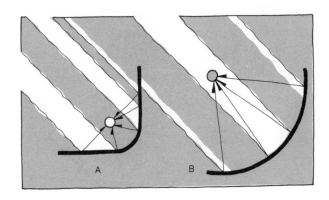

Figure 18B. The difference between figures A and B is in the curvature of the reflecting surfaces. A is what is called a *short-focus collector* because the focal point is close to the reflector. It doesn't need to face the sun as exactly as B, but its more rounded mirror makes it more difficult to build. B, the *long-focus collector,* is easier to make, but harder to operate precisely.

haze or cloud cover. Third, focusing collectors are expensive to buy and complicated to build. (Locating that precise focal point in front of the reflector calls for some tricky mathematics.) Fourth, it's no simple matter to run a stream of liquid through a small focal point to heat it, without affecting the optics of the collector (Figure 18B).

There *are* some collectors on the market that get around some of these problems. The *Sunpak*® solar collector from Owens-Illinois is a series of glass vacuum tubes mounted in front of a flat reflecting surface. In a way, it's sort of a cross between a concentrator and a flat plate collector (see appendix). The Sunpak can absorb both parallel and diffuse light, and can accept radiation from many different angles, which means it doesn't have to move (Figure 19).

Cylindrical collectors are long curved reflectors that intensify sunlight at a ratio as high as 72 to 1. These focus rays on a blackened fluid pipe that passes in front of the collector (Figure 20). They are easier to make than a spherical collector; they can be as long as you like; and they don't need to track the sun so exactly as other types of concentrating collectors.

For the most part, though, focusing collectors have not yet proven practical for heating domestic water in any quantity.

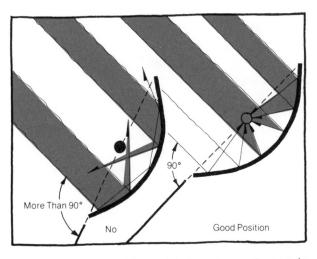

Figure 18A. The problem with focusing collectors is that they rely on direct radiation. They also must always face the sun. When they don't, this is the result. Unless the orientation of the collector is perfect, rays bounce everywhere except against the target.

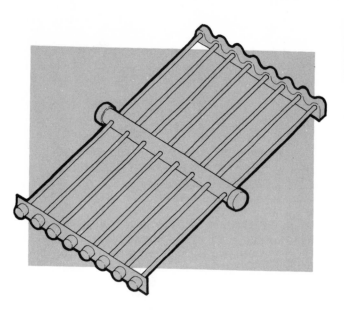

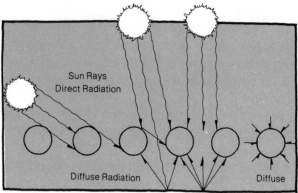

Sun Rays
Direct Radiation

Diffuse Radiation   Diffuse

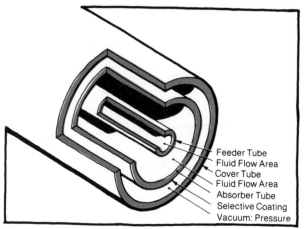

Feeder Tube
Fluid Flow Area
Cover Tube
Fluid Flow Area
Absorber Tube
Selective Coating
Vacuum: Pressure

Figure 19. The *Sunpak*™ solar collector (above) can absorb diffuse light and direct rays from many different angles (top right). Fluid is heated in the absorber tube, and returned through the innermost feeder tube (right). The vacuum space eliminates condensation and holds heat in the absorber area. The absorber tube itself has a special coating that "selects" the right rays to raise the temperature of the fluid.

Figure 20. *Cylindrical* solar collectors have some of the advantages of focusing collectors and some of the advantages of flat plates. They intensify solar radiation, but don't need to track the sun accurately. It is an easy matter, as you can see, to run a channel of fluid through the collector's focal point.

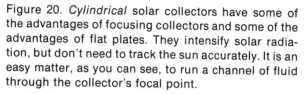

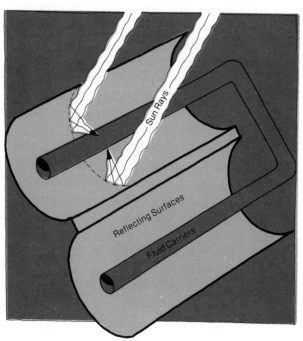

Sun Rays

Reflecting Surfaces

Fluid Carriers

What are Flat Plate Collectors?

Flat plate collectors are much simpler than focusing collectors. They do not need to face *directly* at the sun; they can absorb diffuse light (Figure 21); and almost anyone can make one.

We all know that dark surfaces absorb radiation and that lighter ones tend to reflect it. (This is why desert travelers often wear white clothing and headgear.) A flat plate collector is basically a black sheet of metal with fluid channels or *conduits* running over, under, or even through it (Figure 22).

Have you ever noticed how, on a bright sunny day, a garden hose will squirt hot water after you turn it on? The hose, if it is stretched across the lawn for a while, absorbs some of the sun's rays, which heat the water inside. When you turn on the faucet, this heated water gets forced out of the hose fairly soon, and then the water runs cold.

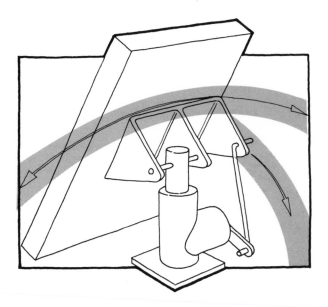

Figure 21. Most focusing collectors must be mounted on a sun-tracking mechanism like this, called a *heliostatic mount.* Flat plate collectors will work somewhat better if they are attached to such a device, but most people think the cost is too high to justify the added efficiency.

Figure 22. The liquid-carrying channels and the absorber plate in a flat plate collector should be of the same material if possible. Tubing can be fastened to the plate in one of several ways, but it's important that there be a good thermal bond between the two. Heat is best conducted to fluid through a plate which has tubing as an integral part of its structure.

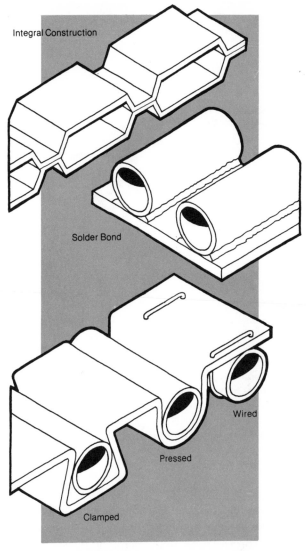

But try turning the faucet down, so just a small stream of water runs through the hose. If you run the water slowly enough, and the sun is hot enough, lukewarm water will dribble out the other end of the hose for as long as the sun is strong. The slower the water is run, and the longer the hose is, the warmer the water will be. That's how flat plate collectors work — only better.

Two things make the collector a more efficient heat trap than the garden hose: (1) The metal collector and its tubes can get much hotter than the plastic hose and the grass beneath it, and (2) the collector plate — sometimes called the *absorber plate* — is covered with glass to hold the heat in (Figure 23).

Flat plate collectors normally operate at temperatures much lower than focusing collectors — somewhere between 100 and 210 degrees Fahrenheit. But they should be able to withstand temperatures a good deal higher than this. Much of the heat that's received by the absorber plate is transferred to the fluid passing through. It might be anywhere from 130 to 210 degrees, depending on how fast it is flowing in the collector channels.

One square foot of collector surface may absorb 700 BTU's of energy or more on a sunny day. A typical flow rate is about one gallon per hour for every square foot of absorber plate. So a 24-square-foot collector in a pumped system should heat at least 24 gallons of water at least one degree in an hour's time — and probably more.

The absorber plate itself is usually surrounded by a metal or wooden box which is enclosed on top by glass or some other form of "glazing."

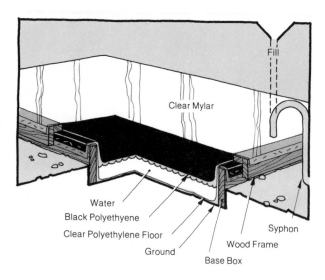

Fill

Clear Mylar

Water
Black Polyethyene
Clear Polyethylene Floor
Ground
Base Box
Wood Frame
Syphon

Figure 23. This inexpensive solar water heater is described by Farrington Daniels in *Direct Use of the Sun's Energy.* The 6-foot by 4-foot box sits in ground which has been dug out slightly. Polyethylene film is laid in the tray to hold water, and a black plastic sheet is laid over the top of the 2-inch deep water layer. A second wooden frame covered with a plastic like Tedlar or Mylar fits snugly over the top of the first. The air space between the upper cover and the black plastic makes a good heat trap. Such a heater can heat 5 gallons of water in about 3 hours. Once it is heated, the water is syphoned out of the box.

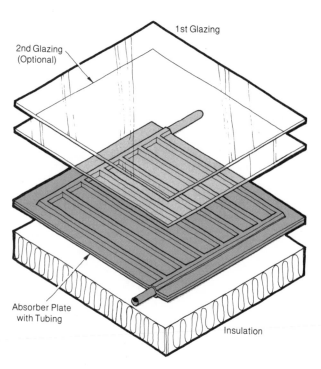

1st Glazing

2nd Glazing
(Optional)

Absorber Plate
with Tubing

Insulation

Figure 24. This is an exploded view of a flat plate collector. The critical element is the absorber plate — in this case the tube-in-sheet type. The absorber is covered with one or more layers of glazing, and rests on a layer of insulation. All of these components are surrounded by a box made of metal or wood.

(This glazing might be fiberglass or plastic instead of glass.) Under the absorber plate there is usually a layer of insulation to keep heat from escaping through the under side of the collector (Figure 24).

The glazing allows visible short wave-length light to pass through to the absorber plate and prevents passing air currents from drawing off much of the collector's heat (Figure 25). But what's even more important is that the glazing keeps some longer wave-length radiation from escaping *back* into the atmosphere. It is this property of glass — and of some types of fiberglass and plastic — that creates a "greenhouse effect" inside the collector.

Everybody knows what it's like to get into a closed car on a sunny day. The heat can be unbearable. Any enclosed space behind glass — whether it be the inside of a car, a greenhouse, or a solar collector — gathers and holds considerable amounts of heat. Short wave-length sunlight comes through the glass and is either absorbed

by, or reflected off walls, floors, car seats . . . or collector plates (Figure 26).

Rays that are reradiated cannot pass back through the glass to the outside because their wave length is made longer once they hit some surface. In other words, a ray that's bounced off an absorber plate has longer wave lengths than it did before it hit. These longer-wave rays reflect off the *inside* of the glass and bounce back and forth inside the air space. In this way a little radiation can become a lot of heat (Figure 27).

So there are four factors that affect the amount of heat captured by a flat plate collector: (1) the solar intensity — "insolation," (2) the collector's orientation toward the sun, (3) the temperature of the air surrounding the collector — call the *ambient* air — and (4) the materials used in the collector itself, specifically the quality of the glazing and the material selected for the key component, the absorber plate (Figure 28).

Absorber plates are most commonly made of copper or aluminum, because these metals do a

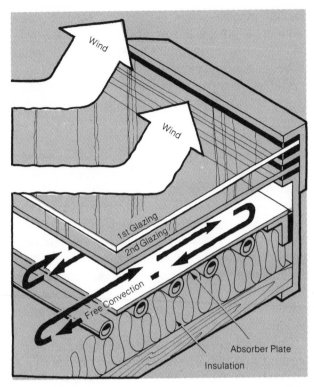

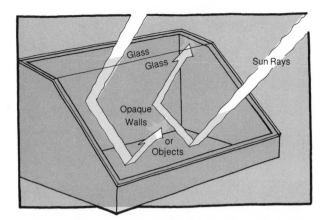

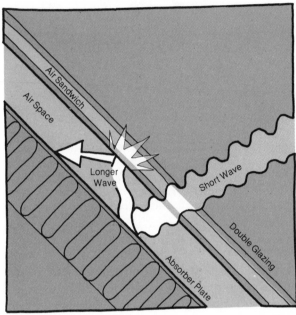

Figure 25. In cross-section a flat plate collector might look like this. The glazing keeps passing air currents from robbing heat from the absorber plate. If a collector is fairly airtight, the small amount of air beneath the glass should be quite hot.

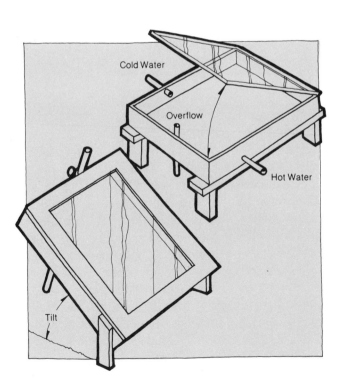

Figure 26-27. A flat plate collector works much like a greenhouse. Rays come through the glass, reflect off walls and the floor of the greenhouse, but can't escape back to the atmosphere. When short wave-length rays hit the absorber plate, some of their energy may be reradiated back, but their intensity is weakened — meaning that the wave lengths are made longer. Because they can't pass back through the glazing, they hit the absorber again and again, giving the plate several chances to absorb them.

Figure 28. Basic solar water heaters like this have been used for more than 30 years in Japan. The lined tray is filled with water in the morning and left in the sun during the day. At night it is emptied and used by the family. In some cases a three-legged box which can be tilted toward the sun is used. Although this is more efficient, it obviously must be made watertight so it doesn't leak when tipped.

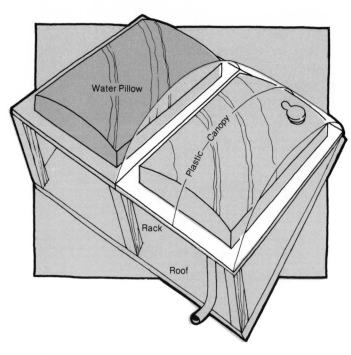

Figure 29. A very inexpensive commercial solar water heater is made by the Hitachi Company of Japan. It is really little more than a black polyvinyl water pillow on a wooden platform. The heater is mounted on a south-facing roof, and holds about 50 gallons of water. In the winter the black pillow is covered with a clear plastic canopy to allow more heat to collect.

better job of conducting heat than, say, steel. (Silver is the best heat-conducting metal of all, but its price obviously makes it out of the question.) Panels with plastic collector plates and no glazing, by the way, are designed only to heat swimming pool water a few degrees, and are in no way suitable for heating domestic hot water. They will disintegrate at the temperatures a domestic hot water heater must reach (Figure 29).

How Efficient Are Flat Plate Collectors?

The easiest way to understand solar collector efficiency is to think in reverse. Don't think so much about the water temperature as about the temperature of the collector itself. The whole system is like a water-cooled engine in a car. When the circulating water — cooled by the car's radiator — keeps the engine running at a reasonably cool temperature, everything is thought to be working efficiently.

The same is true of the solar collector. If water or antifreeze fluid is running through the absorber plate fast enough to keep it cool, most of the heat is being transferred to the water and things are working fine. If the water runs too slowly, and the collector gets hot, there is obviously a lot of *in*efficient heat loss.

In a sense, hot water is just a by-product of the cooling process. The trick then, is to regulate the flow so that *more* water is heated — even though its heated temperature is lower than it would be if the flow rate were slower. When this is the case — the absorber is being kept cool, and the water is hot — the collector is said to be in a healthy state of "equilibrium."

The exact efficiency of a solar collector is figured by comparing the amount of energy that's actually absorbed by the plate to the amount of energy that's falling on the collector. A simple equation would look like this:

$$\text{efficiency} = \frac{\text{energy absorbed}}{\text{energy falling}}$$

If 1200 BTU's of energy a day fall on a square foot of collector surface, and the square foot of the absorber plate absorbs 700 BTU's, the equation becomes:

$$\text{efficiency} = \frac{700 \text{ BTU/ft.}^2}{1200 \text{ BTU/ft.}^2} = 58.3\% \text{ (Fig. 30)}$$

If this seems pretty academic, it is. This is because as a homeowner, you probably have no way accurately to measure BTU's to come up with an exact percentage like this (Graph 3).

Since a collector's efficiency relates directly to its running temperature there are ways to check out quickly how things are going. For example, if the glazing feels very hot to the touch, water may be circulating too slowly. If one collector in an array of panels if overheating — or even if part of a collector is hotter than another part — there may be blockage somewhere in the fluid passages. Usually this will be an air bubble, and air locks like this can be avoided with special plumbing devices.

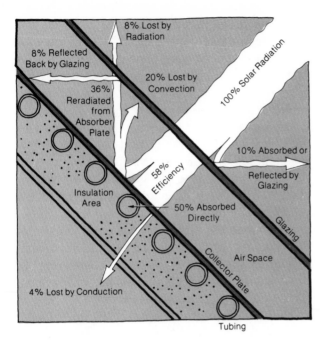

Figure 30. This diagram, after one prepared by Donald Watson, AIA, illustrates why a flat plate collector is not 100% efficient. Only about 58% of the heat actually gets transmitted to the water in the tubing.

A collector can also reach a completely *in*efficient state of equilibrium known as *stasis* when the amount of radiation coming into the collector is equal to the amount of heat being given back to the ambient air. Stasis can occur because of reflection or heat loss from the back because of

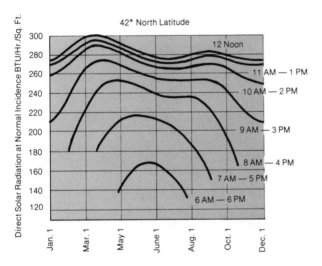

Graph 3. This ASHRAE graph shows the energy per hour that falls on a square foot of collector surface perpendicular to the sun's rays at 42 degrees N.

poor insulation. A clue that this kind of stalemate is in the works would be that the glazing gives off a lot of "glare" during the middle of the day. This means the collector is tilted at a poor angle toward the sun. Sometimes when a collector reaches stasis, it's simply running too hot.

Heat loss from a collector — which spells inefficiency — depends on several factors: (1) how much heat the plate can absorb, (2) how much heat the plate reradiates, (3) the ambient air temperature and wind speed, (4) the number of layers of glazing, (5) how well the glazing lets light through, and (6) how effective the insulation is on the bottom and sides of the collector box. Again, these things are difficult to measure, but they are things you should be aware of as you design your system.

Are There Different Types of Flat Plate Collectors?

All flat plate collectors operate on the same hose-on-the-lawn principle, but there are some variations in how water or fluid flows through them. For this reason, they can be divided into 3 general types: (1) *free circulating collectors,* (2) *forced circulation collectors,* and (3) *open-water collectors.*

In a free circulating collector — part of a "thermosyphon" solar water heating system — the flow of water or fluid through the panel is upward. Here the water movement occurs naturally and is completely self-regulating. The natural flow works because water, like air, rises when it's heated. When the sun shines, the water in the absorber channels gets hotter, becomes less dense, rises through the tubing and moves out through the top of the collector, allowing cooler water to flow in through the bottom opening (Figure 31). (Thermosyphon systems will be discussed in detail in Chapter 8.)

Fluid in a forced circulation system takes the same route through the collector, except that it's pushed up through the panel by some mechanical means — usually a small 1/20 to 1/12 horse-power pump. A system like this can be very

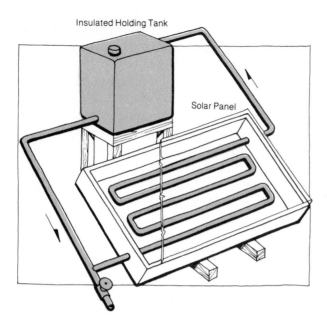

Insulated Holding Tank

Solar Panel

Figure 31. A thermosyphon solar water heater uses a free-circulating panel. Notice how the direction of the fluid flow is upward and that there are no pumps to keep the water moving. This is because hot water naturally rises, while cold water seeks the lowest point in the system.

efficient, but it requires a thermostatic control mechanism to turn the pump off when the sun is too weak to give the collector energy.

The upward flow of liquid through a solar panel can be directed in a couple of different ways. The tubing, which is integral with the plate itself,

might be a single line that zig-zags across the absorber like a serpent. Experts refer to this as a *series pattern*. The advantage of this flow path is that the liquid takes a long route through the collector and gets quite hot. The plumbing involved also is quite simple (Figure 32).

In a *parallel flow* collector, fluid comes into the bottom of the panel through a single in-flow manifold, then is diverted through a number of parallel tubes for one pass across the collector plate, before it's re-collected again in the out-flow manifold (Figure 33). Most prefabricated tube-in-sheet absorber plates are designed to use this flow pattern. Although liquid spends somewhat less time in the collector — and is heated less per pass as a result — more heat will be collected in a day's time.

The third type of collector — the *open-water* type — has an absorber plate made of corrugated steel or aluminum roofing material. Open water collectors are called *tricklers* in the solar trade because water starts at the top of the panel and trickles down the open channels in the corrugations.

The best-known panel of this type is the Thomason collector, developed some years ago in Washington, D.C. The whole Thomason system is well protected by patents, but the collector works very simply. Cold water is pumped to headers along the ridge of the house above the

Figure 32. Tubing in a "series" configuration is easiest to put together and gets water hottest.

Figure 33. A "parallel" tubing configuration, although it's more complex, causes less pressure loss in the system because there's less friction in the pipes.

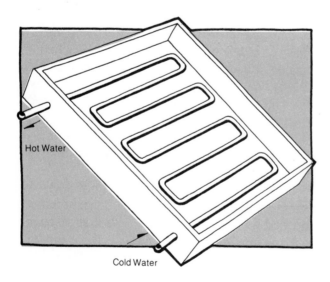

Hot Water

Cold Water

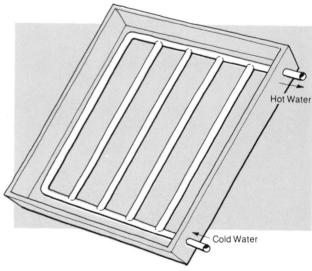

Hot Water

Cold Water

collectors. It then seeps from carefully spaced holes in the header and runs slowly down the black metal absorber plate (Figures 34-35).

At the bottom of the corrugated sheet there is a gutter to collect the water which has taken heat from the sun. This gutter runs into a pipe, which in turn, transports the water to the storage tank. Variations on the open-water collector include the "series dribble" system (Figure 36), which directs water back and forth across the collector plate, and the Shore system, which runs water *between* two sheets of blackened corrugated roofing. There is apt to be lots of evaporation from the open-water in a trickler and this can

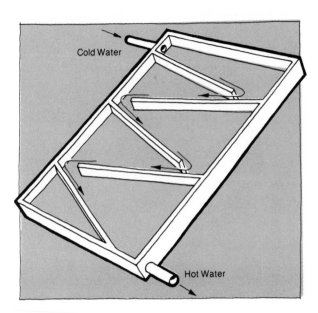

Figure 36. A "series dribble" system is another variation of the open-water solar collector.

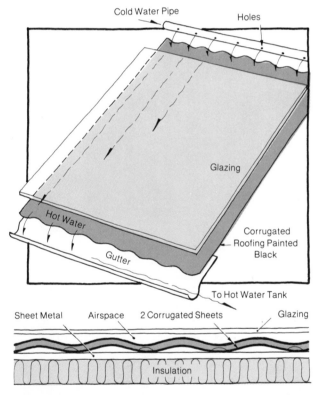

Figures 34-35. Water in a trickling collector, such as the one designed by Thomason (top), starts at the top of the collector and runs to the bottom. The water is heated as it runs down valleys in black corrugated roofing.

In the system designed by Shore (bottom), there are two layers of corrugated metal and the water trickles *between* them.

cause condensation on the underside of the glazing. The Shore collector, with its double layer of roofing, is better in this respect.

Flat plate collectors, for the most part, are either ready-made or built at home. The homemade varieties are built from scratch or put together from a kit. Prefabricated collectors tend to be expensive, as we have already seen. If you're an amateur plumber, you may want to start at ground zero and make collectors entirely on your own. If not, you may opt to buy the collector plate but build the rest of the collector box yourself. It's simple, really. All you have to worry about is keeping the collector box watertight (more on this in Chapter 5). The third option would be to build your collector completely from ready-made parts — available from several manufacturers. (See appendix.)

Whatever you decide to do, build or buy collectors that are small enough and light enough to be handled easily. Remember, there's a big difference between moving a panel on level ground and getting it into place up there on the roof. If the collector is too unwieldy, mounting it properly becomes a balancing act and a wrestling match all rolled into one.

33

CHAPTER 4

Designing Your Collector System

As you get more and more tuned into the sun as a source of energy, you have to consider all the angles — literally. You should be aware of the relative angle of the sun to the horizon — at various times of day and at different times of year. It changes constantly, as you know. You also need to be conscious of how you, your dwelling, your landscaping and your solar collector system stand in relation to the sun. Here the angles become critical.

Spend some time taking an overall look at your home site, and plan the location of your solar collectors accordingly. They will probably be placed in one of three ways: (1) attached to an existing building — the roof of the house, maybe; (2) detached from the building — on the ground, perhaps; or (3) integrated with the building itself. Making the collectors part of the actual structure is tough, unless you build a new house or at least an addition. If you *are* starting a new home, you may want to consult an architect who can help design a house that is in itself a large solar collector (Figure 37).

Here is a brief checklist of things to consider as you plan a new site or examine one you already have:

1. **Building Codes and Zoning Laws.** Do you need a building permit just to install solar panels on your roof? Some places require it. Can you do some selective cutting or pruning of trees on your lot to let in more sunlight? Some forward-looking communities are now zoned with "sun rights," meaning that no one can block the sun's right-of-way to *your* house. Is yours? We're bound to see more of this kind of municipal planning in the future. Are local building codes such that whoever built your house put in a strong rafter system to support the roof *and* the extra weight of solar panels?

2. **Solar Interference.** What obstacles are there that could prevent the sun from shining on your collectors? Can they be removed? Take a long look (and a short look) in every direction sunlight could possibly come from, checking the height of the obstructions against the distance from the collectors (Figure 38). The best time to do this is in December when the sun is at its lowest altitude in the sky. Ideally, the spot where you put collectors should have 6 to 8 hours of possible sunshine every day.

Deciduous (broadleaf) trees some distance from the house may not be such a problem as you think. The sun may pass above them in the summer, and come *through* them in winter

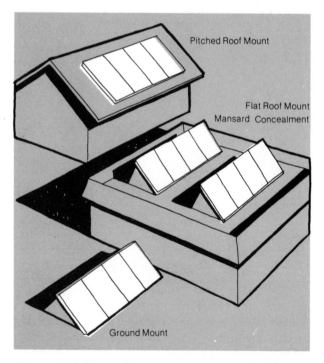

Figure 37. Most flat plate solar collectors are mounted on rooftops. Ground-mounted collectors have certain disadvantages, in that they get covered with snow, are more easily shaded, and can be harmed by vandals. Sometimes a roof is too flat to support collectors faced properly toward the sun. When this is so, they have to be tilted at a better angle, and anchored very firmly so they don't shift in the wind.

34

Figure 38. As you plan your solar system, take a close look at the site. What about mountains, hills, trees and other buildings? Can they interfere with incoming solar radiation? If it's not possible for the sun to shine on your collectors for at least 6 hours a day, a solar system may be hard to justify.

when there's no foliage. Evergreens, however, block more sunshine and cast the heaviest shadow. Some may have to be sacrificed.

3. Determination of True South. Is there a way to mount your collectors to face directly at — or close to — true south? "True" south is marked by the position of the sun at *solar noon* — which shouldn't be confused with noon on the clock. When the sun is halfway between its position at sunrise and its position at sunset, it indicates true south. The easiest way to find true south is to consult a local map for the magnetic variation where you are. More on this shortly.

4. Direction of Prevailing Winds. Will cold winds that blow across your solar collectors steal too much heat? Can they be sheltered from the wind but still be exposed to the sun? Can you build some sort of windscreen to protect your panels?

Heat loss from a single square foot of collector averages 2.5 BTU's per hour when the temperature difference between the inside of the panel

and the outside is just one degree Fahrenheit. That's if the collector has no glazing. With one layer of glass the heat loss is 1.3 BTU's per square foot per hour, and with 2 layers it drops to 0.7 BTU's per square foot per hour. Imagine what heat loss would be if the temperature difference were 120 degrees! And when there is wind, the "chill factor" increases heat loss by almost a degree for every mile an hour of breeze. Keep *that* in mind.

5. Snow Depth. Will snow cover your collectors? Study the roof for a winter to see. Snow slides off roof-mounted solar collectors if the roof is steep enough, but it can drift and settle into the valleys where roofs come together. Collectors that are tucked into shaded corners like this can be obscured from the sunlight (Figure 39).

Figure 39. Snow tends to drift and accumulate in roof valleys, so it's not a good idea to mount solar collectors too close to where two or more roofs join. Study the sunlight patterns on your house at different times of day and at various times of year. Sometimes one part of a house will shade another. Obviously you'll want your collectors in the sunniest possible spot.

Ground-mounted collectors make little sense where it's very cold and there's lots of snow. Not only are they covered much of the time, they're usually far enough from the house that hot water has to travel a long way to reach the storage tank inside. Every inch of distance means potential heat loss — even if the pipes are insulated and buried in the ground. To make matters worse, panels that sit on the gound can be trip-over hazards and attractive targets for rock-throwing vandals.

6. Aesthetics. Is your panel arrangement going to look tacky somehow? Don't forget a basic rule of human nature: If it doesn't "look right" you'll never be satisfied.

Where Should Solar Collectors Be Pointed?

Before orienting your solar collectors, get *yourself* oriented as accurately as you can. Use a compass to find magnetic north, keeping in mind that *it* is not exactly the same as "true" north. To know the "variation" between true north and magnetic north, either look at a recent U.S. map that shows the magnetic variation, or call the United States Coast Guard or the geography department at the nearest university. If you're sure there's no magnetic field nearby — such as a power plant or high-tension electrical lines — that could throw your compass off, make the adjustment for magnetic variation and find true north. True south, naturally, is 180 degrees from true north. In the southern hemisphere, of course, the directions are reversed.

When you get to the point of aligning your solar panels, or designing a new roof to accomodate them, the word is, "Pretty close is close enough." Even if you face your collectors as much as 25 degrees out of perpendicular with a true north-south line, you still get over 90% of the possible radiation (Table 8). In other words, if your roof doesn't face *exactly* the right way, don't panic (Figures 40A, B, & C).

Actually, being off by 15 degrees or so may be

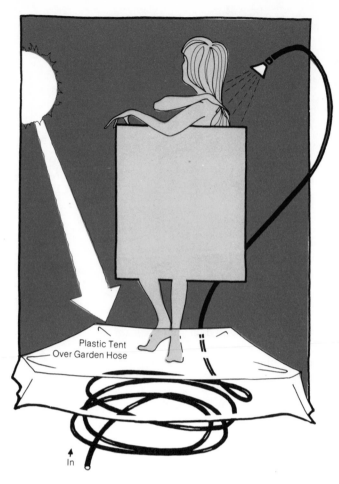

Figure 40A. Simple solar hot water systems are easy to install in cabins and summer cottages, and should cost next to nothing. Some campers have discovered that a coil of black garden hose covered with polyethylene can provide a free hot shower on a sunny day.

an advantage — especially if the deviation is to the south*west* instead of toward the southeast. Morning sun in the east is often hampered by fog or haze, and air surrounding a collector is apt to be 15 degrees warmer in the afternoon when the sun is more westerly. We already know that warmer ambient air means less heat loss.

Make a few trial runs with your collectors before you actually mount them on the roof. Prop them up on the ground at noon, angled so they cast the smallest possible shadow. Then run water through them by temporarily rigging up a garden hose. Later, for a more accurate orientation, make a simple block and dowel device like the one in Figure 41. Advice for mounting collectors permanently on a roof comes up shortly.

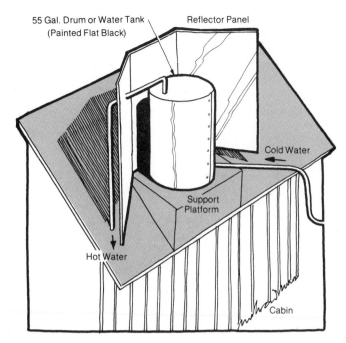

Figure 40B. A 55-gallon oil drum painted black and mounted in front of reflector panels will heat water quite quickly, although it will lose its heat rapidly once the sun stops shining.

Figure 40C. A more efficient — and more elaborate system could be made by laying a used hot water tank on its side against a reflective backing. If it is painted flat black and surrounded with a fiberglass glazing material like Kalwall, the tank should heat and hold hot water all day and well into the evening.

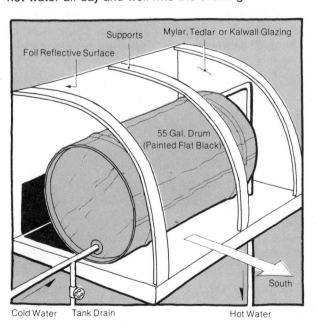

Table 8. As you can see from this table, collector efficiency doesn't fall off dramatically even if the panel is misaligned by as much as 20 degrees. If you have a choice, it is better to point collectors southwest by 15 degrees or so than it is to point them southeast.

How Should a Solar Collector Be Tilted?

Don't confuse orientation with tilt. *Orientation* describes the direction the panel faces, while tilt describes its angle of inclination. *Tilt* is a little more touchy. Ideally the collector should be at right angles — 90 degrees — to incoming solar rays, but this can't happen all the time if the panel stays in a fixed position. After all, the difference between the sun's altitude in June and its altitude in December is 47 degrees. Keep in mind that whenever sunbeams hit a collector at less than a 30-degree angle, the collector reflects off more radiation than it absorbs (Figure 42).

In North America roof pitch seems to get flatter and flatter as you travel further south. This is obviously because houses up north have more snowload in winter. Snow *is* a consideration for

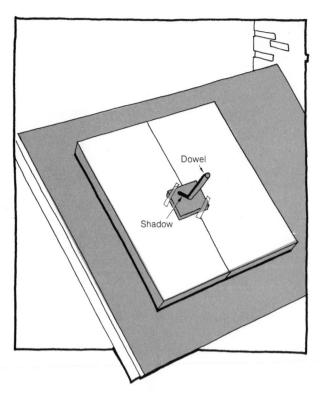

Figure 41. Here is a simple method for setting solar collectors at the best possible orientation and tilt. The target block can be taped against the collector glazing as different positions are tried. To determine the best position for a permanent mounting, this should be done at *solar noon — not noon on the clock*. When the dowel casts no shadow on the target, alignment is as nearly perfect as possible.

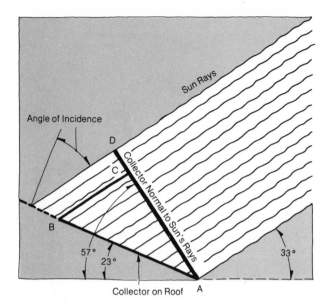

Figure 42. A flat plate collector is most efficient when it intercepts rays at right angles. Once it's mounted in place it's impossible to keep the collector perpendicular to incoming rays at all times. So a compromise position has to be found. The angle between the radiation line and a line perpendicular to the collector is called the *angle of incidence*. As the angle of incidence increases, the collector's efficiency decreases.

solar collectors, as we already know. If snow is to avalanche off a collector consistently, it must be tilted not less than 40 degrees from the horizontal.

An informal rule among architects is that roof pitch should be equivalent to the latitude of the homesite *minus* 10 degrees. (If you live at a latitude of 35 degrees North, your house might have a roof pitch of 25 degrees — although this formula is *not* hard and fast, especially with more modern homes.) Determine the *actual* roof pitch of your house either by going out on the roof with a long carpenter's level and a large protractor, or by referring back to the original blueprints to look up the roof pitch. (Table 9 will help you convert pitch in inches per foot — the indication on most blueprints — to degrees of steepness.)

Most authorities agree that ideal collector pitch for solar space heating — *not* hot water heating

— is the latitude of the home site *plus* 10 to 15 degrees. In other words, if you live at 35 degrees N. your collectors should be tilted at 45 or 50 degrees for winter solar space heating. (You can find your latitude, by the way, by looking at any road map.) As you go farther north, roofs need to be almost vertical. What all of this means is that the normal roof pitch is about 20 degrees off of the ideal tilt for maximum wintertime solar collection.

Don't be disheartened. For heating water only — not living space, now — a solar collector tilt *equal* to your latitude is just about right, since we're trying to make hot water in summer as *well* as winter. To say it another way, solar water heating calls for a flatter tilt. What you *do* have to decide is whether you want to leave your collectors flush with the roof — at a slightly wrong angle for optimum efficiency, perhaps — or to make tilt adjustments. Experience has shown that it's better in most cases to leave collectors flat on the roof, even if it means more panels.

If you lift the top end of the panel many degrees off the roof you are leaving yourself open to

Angle in Degrees	Conversion of Pitch to Degrees Pitch in Inches Per Foot
10	2.1
15	3.2
20	4.4
25	5.6
30	6.9
35	8.4
40	10.1
45	12.0

Table 9. Architects, contractors and solar engineers sometimes express roof pitch differently. Instead of stating the pitch in degrees they often think in terms of inches per foot. A three-on-twelve pitch means that the roof rises 3 inches for every 12 inches of horizontal distance. This table can help you make conversions from one to the other. A 2-on-12 pitch is roughly equivalent to 10 degrees of steepness, while a 12-on-12 pitch is equal to 45 degrees. On some solar homes roof pitch may be greater than 45 degrees.

possible roof leakage and wind damage. A collector that gets mounted on a "prop" can act like a sail, exerting tremendous strain on itself, the roof, and its mounting. It would be no fun at all to come out your door one morning and find your expensive solar collectors shattered across the front lawn (Figure 43).

The best compromise, if your roof is anywhere near the correct pitch, is to mount the collector panels flat, and either add more collectors or solar reflectors. In short, if your roof is much too flat, think about making tilt adjustments *only* if you're willing to build an exceptionally strong supporting structure for the panels.

How Much Do Solar Reflectors Help?

Reflector panels which bounce rays back onto a collector should reduce the amount of collector surface you need. In fact, a small collector can receive more radiation than a much bigger one if it's helped by properly arranged reflection surfaces. Good reflectors should boost the performance of *any* flat plate collector by at least 25% (Figure 44).

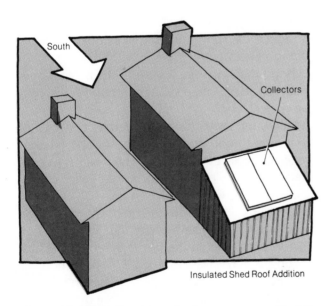

Figure 43. When the ridge of a house runs north and south, it can be difficult to design a collector system that works well without having to use many panels on both the east and west roofs. A cheaper alternative might be to plan a simple addition with a shed roof that puts the collectors at just the right tilt.

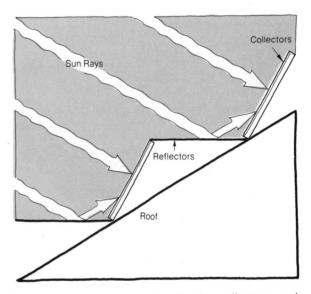

Figure 44. Sometimes banks of solar collectors and reflector panels can be made to work in combination. Reflectors can increase the performance of flat plate collectors by 25% or more. It's best, of course, if the reflector is adjustable, so it can be moved to shed light on the collector at all times of the year.

To work best, reflectors should be mounted *above* collectors in winter — so they receive rays from a low angle and throw them back onto the absorber plate — and *below* collectors in summer when the radiation is coming from a much higher angle.

Shiny aluminum mounted on plywood makes a fine reflector, but aluminum is fairly expensive and not easy to stick to wood. Glossy white enamel paint will work nearly as well. Best of all are aluminized Mylar sheets that reflect beautifully and can be glued to plywood.

If you decide to make moveable reflectors, build them to be sturdy. *Over*build them, in fact. They're going to stick up in the wind and will need to be braced every which-way. If you hinge the top or the bottom of the reflector panel, buy the biggest, heaviest strap hinges you can find (Figure 45).

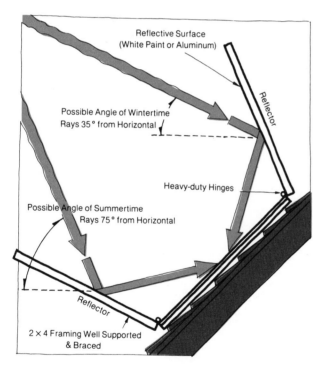

Figure 45. Light comes off a reflector surface at exactly the same angle that it hits. In winter radiation comes from a lower angle than it does in the summer, so ideally a collector should have reflectors both above and below it. Reflector panels are apt to be very exposed and can take a tremendous beating from the wind. They should be well built, braced and guyed.

When you get ready to adjust the angle of the reflector in relation to the collector, do it sometime around noon. Remember that rays will reflect off it at an angle equal to the angle of the incoming rays. In other words, if radiation strikes the reflector at a 45-degree angle, it will also come off at a 45-degree angle, meaning that the reflector and the collector should be in a 90-degree relationship. (Usually a relationship in the range of 90 to 105 degrees is acceptable — it doesn't have to be exact.)

And remember what was said a few pages back: if solar rays hit a collector panel at less than 30 degrees, most of the energy is being reflected off the collector glazing and wasted. Moveable reflectors then, can insure that sunlight hits the collectors at a favorable angle all year long. Because there's no delicate plumbing attached to the reflector, *it* is a whole lot easier to adjust than the collector itself.

Some reflector panels are designed to do double duty as insulating covers that close over the collectors at night and on cold cloudy days. These can be made with styrofoam or urethane insulation, fastened to a plywood backing that is framed with 2 by 4's. (The waterproof adhesive that's used to stick modern wall panelling to old walls works great for fastening foam insulation to plywood, by the way. It comes in a tube that fits in a regular caulking gun, and is sold under many brand names.)

Keep half an eye out for other natural or man-made reflecting surfaces around you. It may be possible to get reflected radiation off a neighboring metal roof or a nearby pond, lake or swimming pool. And snow may be the greatest reflector of all, even though winter seems to contradict the whole idea of increased collector efficiency.

Dr. Harry Thomason, the man who developed a solar heating system using the "trickling" collector mentioned in the last chapter, tells of how astonished he was one cold snowy morning shortly after his solar home in Washington, D.C. was finished. He called his wife from the office for a routine check on how the system was working. To his surprise she reported that water coming off the collectors was several degrees hotter than it had been the day before — when the temperature of the outside air was a good deal warmer.

He thought she must be mistaken, and asked her to check again. She came back on the phone with the same report. He concluded that the collectors were generating more heat because more photons were hitting them after reflecting off the snow. He was right.

How Should Solar Collectors Be Mounted?

If you plan to build your own collectors, you may want to go on to the next chapter first and then come back to this section. But before getting too specific about construction details, let's discuss some general things to consider when mounting solar collectors on a roof.

It's obviously important to know the size of the collectors you're going to mount. They'll probably be somewhere between 24 and 36 square feet in size. It's also important to know what they're all going to weigh. To be on the safe side, figure that a typical solar collector weighs between 2.5 and 3 pounds per square foot. (If a collector has two layers of glazing — and they're both *glass* — it could weigh somewhat more than this.) That means that a 24-square-foot panel with just one glazing — say 3 feet by 8 feet — will weigh about 75 pounds. A panel that is 32 feet square should weigh about 100 pounds. (Glazing, by the way, especially if you use glass, should be installed only *after* everything else is in place on the roof.)

Once you have figured the total weight of your collectors — by multiplying their total square footage by 3 pounds — take a quick look at your rafter or truss structure. Most rafters in northern areas are more than adequate to support the roof and the additional weight of solar collectors and a heavy snow load. But if you have *any* doubt, check with a local carpenter, contractor or building inspector.

All he'll need to know is the length and size of the rafter beams (2 × 4, 2 × 6, 2 × 8, 2 × 10's or whatever), how far apart the rafters are spaced (12 inches, 16 inches, 24 inches, etc.), the total weight of the collectors, and the pitch and nature of the existing roof (shingles, metal, slate and so on). The inspector can tell you — possibly over the phone — if and how much additional reinforcement you'll need.

What he's most likely to say is that the weight of the collectors is even lighter than a moderate to heavy snowload on your roof, and not to worry. If he does express some concern, making the roof strong enough may be just a matter of putting in some extra support posts and tie beams, and "sistering" the rafters or trusses — beefing them up with some additional lumber.

In some cases where a system is going to be installed at a very northerly latitude, collectors may be mounted vertically on a wall to be assisted by reflection from snow. Wall-mounted collectors can be quite efficient in wintertime, but almost worthless in summer unless they're rigged with a reflector. They should be fastened with enough support so their weight puts no strain on the piping that's coming and going from them. In other words, don't try to hang them by their own plumbing (Figure 46).

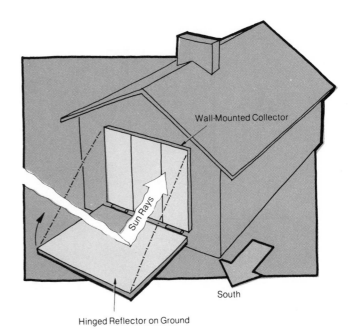

Figure 46. Some wall-mounted collectors work well in northern latitudes in winter. They accept the sunlight from a low altitude during the coldest months, but may need a reflector that can be elevated in summer if they are to receive enough radiation from a higher angle. Insulated reflectors can be designed to fold over the collector panels to protect them from cold and driving winds.

As a matter of fact, *any* collector should be mounted as solidly as possible, so it has no chance to shift and snap a plumbing connection somewhere. This may mean toenailing through the collector box and into the roof, screwing the panels in place, or using heavy galvanized lag screws and angle brackets to anchor the panel. Either way, don't be timid about making the mounting as strong as you can (Figure 47).

You'll also need to make sure the header pipes running to and from the collectors are absolutely level — as are the tops of the panels themselves. Otherwise there will not be even water flow through all the water channels in the collectors. This may call for some precision shimming in

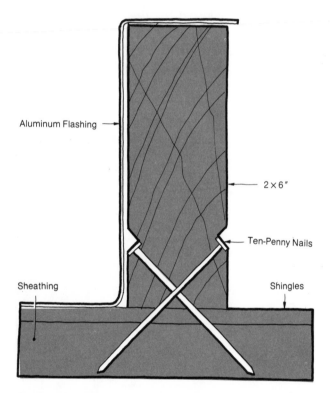

Figure 47. The collector box — often made of 2x6's — is sometimes toenailed to the roof sheathing with 10-penny or even 16-penny galvanized nails. Water must be kept out of the collector box so its top and sides should be protected by aluminum flashing. The outside of the wooden box itself should be painted with a high-quality exterior paint or wood preservative. Don't paint the inside because the paint may react poorly to heat, vaporize and fog the inside of the cover glazing.

Labels on figure: Aluminum Flashing; 2 × 6″; Ten-Penny Nails; Sheathing; Shingles

places, so bring along your level and maybe a few tapered cedar shingles. They shim better than anything else, as any carpenter will testify.

Once the collectors are installed and fastened to a "fare-thee-well," they'll have to be made leakproof by adding bendable aluminum sheeting called *flashing*. Begin flashing at the bottom of the collectors, work up along the sides, and then across the top. Be sure there's generous overlap at all the flashing joints. The upper piece always laps *over* the lower piece (Figure 48).

Tuck the top piece of flashing under the first course of shingles (or other roofing material) above the top of the collectors. Any roofing nails that are driven through the aluminum flashing — and you should try to use as few nails as possible — should be smeared with asphalt roofing cement so water can't seep in past the nail heads.

People sometimes ask if one type of roofing or roofing color is better than another around solar collectors. Some experts claim that a shiny metal roof reflects some light and encourages snow to slide off more easily. Others argue that a black asphalt paper or black shingled roof absorbs more radiation, melts snow, and raises the temperature of the ambient air near the collector on a very still day. When you come right down to it, though, surrounding roof color and texture have very little effect on a collector's performance.

Glass on collectors *can* be protected from hail, falling branches and rocks if you lay hardware cloth over the top. But avoid this if you can. It will reduce the collector's performance by about 7%.

How you actually organize the collectors on the roof will depend a lot on their size, how many you have, how big the roof is, what is the most convenient way to make plumbing connections, and whatever obstructions there may be up there. For example, you may have to work around chimneys, vent pipes, stove pipes, dormers, and maybe even a TV antenna. These obstacles will have an effect on how you arrange the plumbing to move the fluid to and from the collectors.

If your roof is simple and doesn't demand a fancy layout, you can take the most obvious approach — *end feeding*. In this case water goes in one end of the lower header and out the opposite end of the *upper* header pipe. Again, with this type of installation it's vital that the lower header

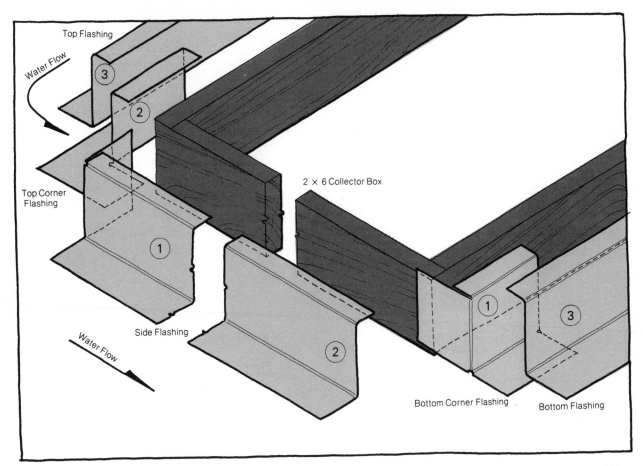

Top Flashing

Water Flow

Top Corner Flashing

Side Flashing

Water Flow

2 × 6 Collector Box

Bottom Corner Flashing

Bottom Flashing

Figure 48. Installing flashing neatly and well is a tricky process. The aluminum material is easy to cut with tinsnips, but it will bend, dent and kink in all the wrong places unless you work with it carefully. Here is how to fit flashing around the top and bottom corners of the collector box. Remember that a top piece must always lap over a lower one. If you leave plenty of overlap, you can't go wrong.

be horizontal. If the headers are going to serve several panels in a line, they must, of course, be *straight* as well. Little dips and humps in these pipes can cause all kinds of trouble as time goes on.

The other option is to use *center tee feeding*. Here water is fed to the roof collectors at the center — between panels on either side. A "tee" fitting divides the flow of water to both sets of panels. The problem that can develop is that you get an uneven flow of water if the tee somehow diverts more water to one side than the other. (The tee is not a regulator, remember.) (Figure 49, 49A.)

To know if there is uneven water flow watch the temperature of the collectors on either side of the divide. When one gets a lot warmer to the touch than the other, it's not getting as much water circulation. This dilemma is easy to prevent if you use a little foresight. Install a *gate valve* — like an in-line faucet — on either side of the tee, so you can regulate the amount of water flow. Turn one or both up or down to keep an equal amount of water feeding to either side.

One last important thought before leaving this subject for something else: Leaving a collector stagnant in the sun — with no fluid running through it — is a sure way to destroy an expensive piece of equipment. When you get your collectors installed and glazed, cover them with a tarp or some scrap pieces of plywood until they are fully plumbed, connected to the storage tank, and ready to operate. It'll save lots of grief later on.

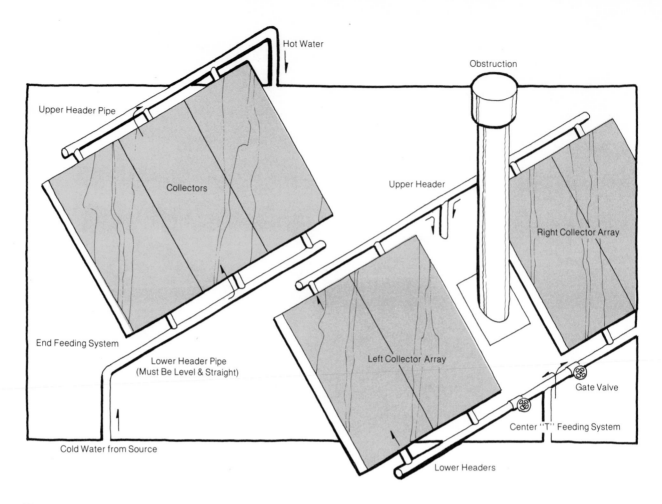

Figure 49. These are the two most typical header arrangements. End feeding brings water into the collectors at one side and out the other. Another arrangement splits water at a central "tee", sending it to panel arrays on either side. Some valves may have to be installed to keep the water flow equal on both sides. In cold climates, of course, the plumbing may not be exposed as it appears in this illustration.

44

INSTALLING
COLLECTOR PANELS

This procedure is a little more intricate than ordinary mounting methods. This is because the top and bottom header boxes have outside pieces that require a bevel cut, which should be done on a table saw. Even so, the end result is neat, simple-looking, and relatively trouble-free.

It will work on either new or old houses. If it's a new house, a layer of waterproof roofing — asphalt roofing paper, for example — should be laid on the roof sheathing first. Only a few inches of overlap are necessary, rather than the traditional half lap. The roof beneath the solar collectors must of course be waterproof, but it won't have to stand up to extreme weather exposure. On an old house that's being "retrofitted" to solar collectors, the panels can be laid right over the old shingles or other roofing.

Before you start, you'll need a set of carpenter's staging brackets so you can work safely on the roof. These are common items that you should be able to rent locally.

Figure 50

1. The first step is to lay out the complete collector array on the roof, using a carpenter's chalk line as in Figure 50. This is to get everything centered, and to be sure that plans you've sketched on paper will work out in practice. On older houses the roof surface may not be perfectly true. If this is the case, the collectors may have to be shimmed to get them lined up properly.

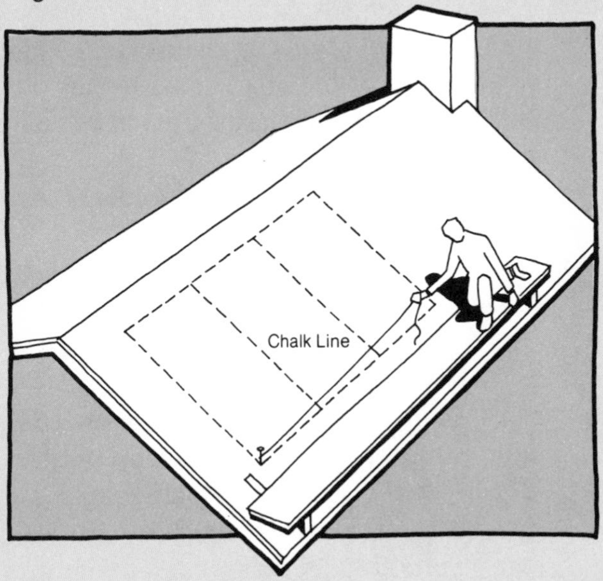

2. Now you must decide how you're going to install the collector headers. Will they run above the roof, or will each panel be connected to an interior manifold running inside the attic under the roof sheathing? The first means few holes that have to be sealed and caulked. The second is easier to put together. The next few steps de-

scribe on-the-roof header construction in detail, but they also apply to collectors that are plumbed individually through the roof.

45

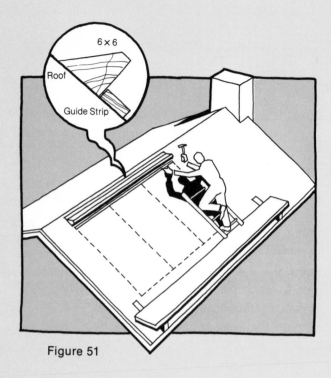

Figure 51

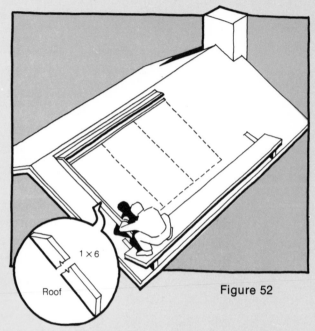

Figure 52

3. Build the box that will house the top collector header, using a beveled 6 x 6 as shown in Figure 51. Screw it, along with the guide strip, to the roof with rustproof screws, making sure that both the box and the strip are perfectly horizontal. The plywood top of the box should be left off until the panels and headers are in place.

4. Now add the left-side trim board as shown in Figure 52. This is just a 1 x 6 angled at either end to match the bevel cut on the top and bottom pieces.

5. A row of panels can now be laid in place. Start at the left side and work to the right, just laying the panels loosely on the roof. Next they should be butted against the guide strip, straightened and leveled (see Figure 53). A strip of wood can be tacked in place along the bottom of the panels to keep them from sliding down the roof. The individual boxes can be toe-nailed

into the roof sheathing with 16-penny galvanized nails, leaving 1/8 inch of space between collectors.

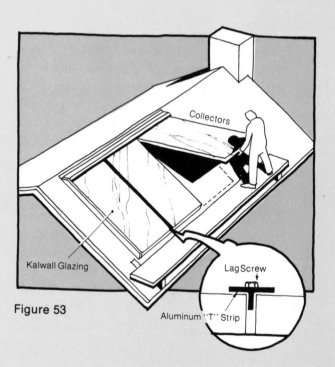

Figure 53

46

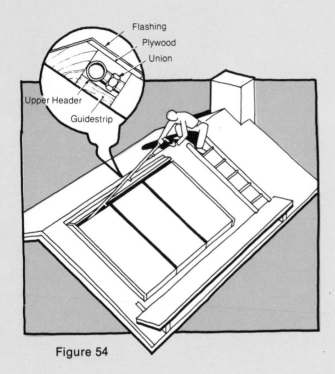

Flashing
Plywood
Union
Upper Header
Guidestrip

Figure 54

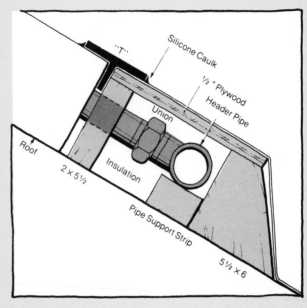

"T"
Silicone Caulk
½ " Plywood
Header Pipe
Union
Roof
Insulation
2 x 5½
Pipe Support Strip
5½ x 6

Figure 54A

6. Next, the panels can be further fastened in place using aluminum T-sections and long, gasketed lag screws. These will need to be at least 7 inches long. Run a double bead of phenoseal caulking down the edge of each collector box before the T-sections are screwed down. This caulking should seal the joints.

7. Add the right-hand trim piece.

8. Assemble the top header out of copper tubing. Solder all the joints carefully and test for leaks. All of this can be done on the ground. Connect the header to the panels on the roof with a union. See Figure 54.

9. Screw the plywood top of the header box and finish the flashing as shown in Figure 54A. Stuff the box with fiberglass insulation before you close it.

10. Build the bottom header box, fasten it to the roof horizontally, and build the insulated bottom header. Install them both, and stuff the box with fiberglass insulation — once you've made sure the header itself is level and straight. See Figure 54B.

11. Double check all lag screws and other fasteners to make sure the panels are held firmly in place. Caulk all joints thoroughly.

47

CHAPTER 5

Building Solar Collectors

If you can read a measuring tape, make a square saw cut, drive nails or screws, drill holes, and use a paint brush, you can build your own solar collectors — *provided* you buy the absorber plate itself. Making your own absorber gets fairly complicated, and hardly seems worthwhile since there are several fine prefabricated plates on the market right now. Besides, you probably won't save much money building your own.

Before you get started, there are four things to think about:

1. Select (or build) the collector plate carefully. It may be the most critical element in the whole system.

2. Use galvanized nails and other waterproof fasteners in the collector box. Your panels will last many times longer if the hardware can't corrode.

3. Learn to use a caulking gun if you don't already know how, and don't be afraid to use caulking liberally. Your collector will need to be waterproof.

4. Use plenty of paint or wood preservative on the exterior wood parts of your collector. The longer you make *them* last, the longer your panel will survive. Well-treated wood should last a lifetime. But don't paint the *inside* of the collector box. Paint, creosote or Cuprinol will only vaporize in the intense heat and condense on the inside of the glazing.

How Do You Judge the Worth of an Absorber Plate?

Collector plates are usually made out of one of three materials: aluminum, copper, or steel. One of the ways to judge how well an absorber will perform is to examine the "thermal bond." Heat is transferred from the plate to the water by *conduction*. In other words, the best exchange is made because the two materials — fluid and metal — are in direct contact with each other.

Heat can flow from a warm material to a cold material in any direction — up, down or sideways. In a collector other than a trickler it doesn't matter if the tubing in the absorber is above, below or within the collector plate itself. What *does* matter is that the tubing be bonded to the plate as closely as possible, so that direct contact is made along its entire length. To say it another way, a tube that's simply spot-soldered, tack-welded or wired to the collector plate will not have as good a thermal bond as a tube that's soldered all the way along. When it's soldered this way it's said to be "filleted." (See Figure 55.) Filleting is just one thing to look for.

Another thing to examine is the material. Aluminum is cheaper than copper, but it doesn't conduct heat nearly so well. That is, copper doesn't need to absorb as much heat before it begins to transfer some of its warmth to the fluid in the channels. In areas where there is even a remote chance of freezing — practically everywhere in the continental United States — a mixture of water and non-toxic antifreeze is normally run through the collector panels, as we'll see later. Most kinds of antifreeze react poorly with aluminum, causing the plate to corrode much more quickly than a copper one. Copper is corrosion-resistant.

You also have to be a bit wary of steel. Hot water *without* anti-freeze corrodes steel very quickly, although chemicals in the antifreeze will allow the plate to last somewhat longer. Whatever the plate material is going to be, it should be the same as the tubing. They both must expand and

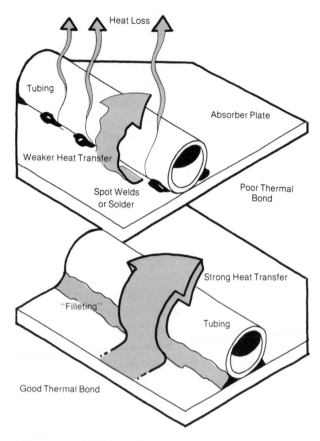

Figure 55. Solar radiation is changed to heat at the blackened absorber plate. The heat is then transferred to the fluid flowing inside the tubes. Here is where a good *thermal bond* is critical. Solder or welding should run along the entire length of the tubing so there's direct contact all the way. If not, heat will escape.

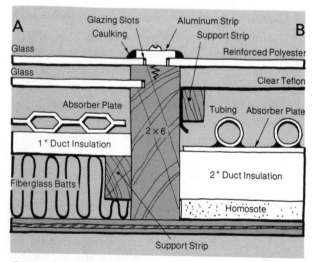

Figure 56. Keep referring to this sketch as you read Chapter 5. It shows several options for constructing a flat plate solar collector.

contract at the same rate, or the bond between them will be strained and eventually break.

It's bad enough that steel can only be used in a "heat exchanger" system where there's an antifreeze solution. To make matters worse, you can't solder copper tubing to a steel collector plate because copper expands so much more than steel, and if you try to fasten galvanized pipe to a steel absorber plate by welding the two together, you won't get a good weld — even though you may get what welders call the "zinc shakes" from breathing the fumes while you try.

Besides, most of the plumbing that transports the antifreeze mixture to and from the storage tank will probably be copper tubing, and connecting galvanized pipe to copper pipe means you have to buy and install connector fittings. If the plate and tubing are already copper, you just need a simple soldered joint. *In short, a copper absorber plate is by far your best investment.*

Some collector plates, like the one made by the Olin Company, are made with a "roll-bond" process. This procedure sneaks the liquid-carrying channels right into the plate itself, making the absorber an even more efficient heat transmitter (Figure 56, section "A"). At Olin, the tubing pattern is laid out on one sheet of metal — in this case aluminum — and then a sandwich is made with a second sheet of aluminum. The two sheets are bonded together everywhere *except* where the channels are going to be.

A special needle is stuck into the unbonded

Side A shows two layers of glass. Notice the spacing between them, and the spacing between the second layer and the absorber plate. The top cover rests on the top of the collector box. Its edge is held with an aluminum strip and sealed with caulking. The second layer of glass fits into a slot for support. In this case the absorber is a prefabricated plate manufactured with a "rollbond" process. Rigid duct insulation supports the absorber.

The bottom of the box is plywood, and the sides are regular 2x6's.

Side B shows an outer glazing of reinforced polyester and an inside glazing of Teflon plastic film. The film can be stapled to the inside of the support strip before the strip is nailed to the 2x6. Here the absorber is copper with copper tubing. It sits on a 2-inch layer of duct insulation and a layer of Homosote — also a fine insulation.

area and fluid is injected under tremendous pressure. This great pressure forces the sheets apart where they're not stuck together, forming the tubes where liquid can flow. Other companies get similar results with other materials.

Prefabricated absorber plates can run as high as $2.50 to $10.00 a square foot, and you have to shop cautiously because standards for commercial manufacturers are not yet fixed. The *best* absorber plate we know of at this writing is the *Kennecott Terralite copper absorber* made in Lexington, Massachusetts (see appendix).

Aside from being efficient, the Kennecott is cheaper than anything you could build from scratch. The plate itself, including the tubing, is very thin — about 1/4 inch — but totally adequate. It withstands temperatures all the way from −30 degrees Fahrenheit to +300 degrees, and comes already precoated with flat black paint.

What's most attractive about the Kennecott absorber is its price — $2 to $3.00 a square foot, all plumbed and ready to go. At this rate a 3-foot by 8-foot plate will run about $60. Buying a premade absorber at this price is by far the easiest and cheapest way to build a solar collector.

How Can You Build Your Own Absorber Plate?

Some ambitious and highly-skilled types will surely choose to build their own absorbers. Several good plans have been published lately, and even though they require a lot of picky labor, the final products work pretty well.

One detailed set of instructions is put out by the Brace Research Institute near Montreal, the world-renowned center for studying intermediate technology. Their plans call for a sheet of galvanized corrugated steel roofing riveted to a flat sheet that is slightly larger. The sides of the larger bottom sheet are crimped around the sides of the corrugated top sheet, and the ends are designed to accept short sections of galvanized pipe at either end. These pipes feed and drain the water

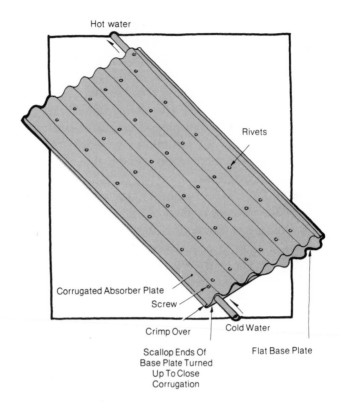

Figure 57. The drawing is taken from plans drawn up by the Brace Research Institute in Canada. The flat sheet is riveted to the corrugated top sheet, and the rivets are then soldered to prevent leaks. The end cuts can be made quite accurately if you make a cardboard pattern and trace it onto the metal before you cut. The problem with this homemade absorber is that water doesn't always flow through it evenly.

channels between the two metal surfaces (Figure 57).

There are several other ways to build absorber plates with corrugated metal to make trickling collectors à la Thomason and Shore. Water can run either openly in the exposed corrugations or between the sheets (Figure 58).

Both the Florida Energy Committee and the Langley Research Center in Hampton, Virginia have published excellent pamphlets on how to build solar collectors. The one from Florida is called *How to Build a Solar Water Heater* and costs $3.49. The Langley book, published by the U.S. Department of Commerce, is entitled *An Inexpensive Solar Heating System for Homes.* Its price is $4.25 (see bibliography).

Each of these helpful booklets describes ways to bend 3/4-inch copper pipe into a series con-

51

figuration across a copper sheet. The problem immediately apparent here is that bending pipe evenly for almost 180 degrees requires special skills and equipment. And bonding it to the copper sheet is a painstaking process that could only be done by someone with plenty of skill and experience. (It also looks like it would take miles of solder.) So it's not something we would recommend to beginners, even though the directions *are* very complete.

If you're a skilled plumber, and you do choose to make your own absorber plate using a series fluid-channel configuration, double the tubing back and forth across the plate leaving 4 to 6 inches between the runs. The temptation will be to space the channels closer than this to increase the distance fluid must travel through the panel.

But studies show this only creates a bottleneck in the water flow. The absorber will *not* be more efficient with more closely spaced tubes. That only makes more friction in the pipes (Figure 59).

If you decide to make an absorber plate with a parallel configuration of fluid channels, the spacing between pipes still should be about 5 inches. The parallel riser tubes will want to be 3/8- to 1/2-inch tubing, while the inlet and outlet manifold pipes may have to be 3/4 inch in diameter to handle the load. (Risers in a thermosyphoning collector should be larger than 1/2 inch.)

The advantage of parallel channels is that less friction is created in the collector. The problem is making leak-proof connections between the risers and the manifolds because they're of different sizes. This either means many expensive fittings if

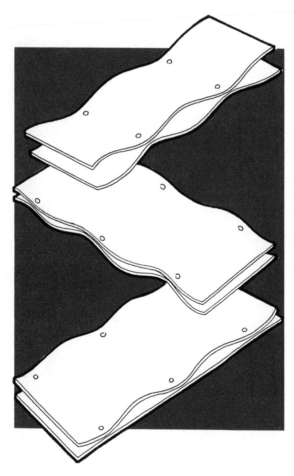

Figure 58. Do-it-yourself absorber plates are frequently made of corrugated metal. As you can see, parallel water channels can be created in several different ways.

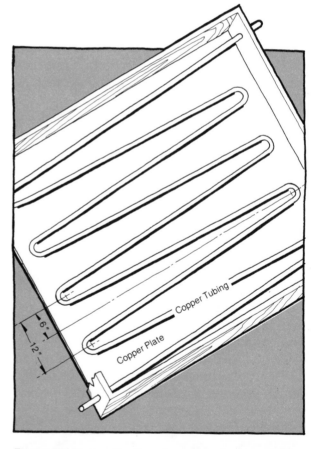

Figure 59. If you plan to run a series of copper tubing across a copper plate to make your own absorber, you'll want to lay it out in a pattern something like this. The tubing should not double back on itself to be any closer than 6 inches.

you use copper, or some exact drilling and welding if you use steel or galvanized pipe. Dissimilar materials *cannot* be used. If you try, you'll only end up with "galvanic" corrosion at the joints. In any case, a parallel system of tubes should be assembled and thoroughly pressure tested before it's bonded to the plate itself.

What About Painting the Absorber Plate?

The absorber plate needs to be black to absorb much solar radiation, as you already know. There are special selective paints developed specifically for solar systems. These are expensive *because* they're so special, but they allow short wave radiation to be absorbed by the plate, and at the same time retard *re*radiation of longer wave-length rays. Unfortunately some of these selective paints can't be easily applied at home, so we have to find substitutes.

Shiny black paint, of course, reflects too many rays. And, some normal flat black enamels will not hold up well at the temperatures a solar collector has to bear. They will crack, peel and chip, even if they're properly primed and applied to a clean, dry surface as the manufacturer recommends.

One alternative that's sometimes suggested is the *anti-glare paint* that's sprayed on aircraft right in front of the windshield. But this isn't the easiest thing in the world to find, naturally. Sears driveway coating, of all things, works pretty well if you use 2 coats and follow the application directions to the letter.

Sherwin Williams, Rustoleum, and others make high-carbon, flat black paints that are more than satisfactory for absorber plate coatings. Two small cans will cover a 3-foot by 8-foot absorber plate easily. The very best paint — though it's somewhat expensive — is *3M Nextel Black Velvet,* which should be applied over Nextel gray primer. This is recommended above all others.

Let's summarize what's been said so far: our best advice is to buy a precoated copper absorber plate and build your own collector box.

How Do You Build the Collector Box?

Anyone who can avoid hitting his thumb with a hammer can build a collector box. Usually it's nothing more than a simple frame made of 2 × 6's. Doing it's like making a kid's sandbox, except that sometimes the wooden frame is covered with 1/8-inch aluminum angle pieces that are mitered at the corners. This isn't hard to do either.

All you have to worry about is making the box as waterproof and dust-proof as you can. Any condensation that's allowed to build up on the glazing will cut down on the collector's efficiency. So will dust. If *it* collects on the collector plate it will absorb a lot of heat but have no way to conduct this heat to the liquid in the channels. So wetness and dust are the main reasons for using plenty of caulking and doing a careful job of flashing.

It's best to use redwood, cedar or cypress 2 × 6's, since spruce and pine will rot more quickly. It makes more sense, by the way, to use 2 × 6's than 2 × 4's because they are wide enough to allow at least an inch of space between the outer glazing and the absorber plate, plus room for insulation beneath the plate. If the sides of the box are much *higher* than 6 inches, the absorber plate may be unnecessarily shaded when rays are hitting the box at a low angle (Figure 60).

The bottom of the box can be 3/8-inch or 1/2-inch plywood, and the plate itself can be supported by pieces of firring strip (often called *strapping*), dowels, wires, or rigid insulation below the plate. The second layer of glazing (if you decide to use one) can either be fastened to a strip of wood at its edges or fit into slots that are rabbeted into the sides of the collector box. These slots should be cut with a table saw before the box is put together — if you hadn't already figured that out. And don't forget the holes for the inlet and outtake pipes. These should be as small as possible and sealed with caulking.

The amount of airspace between the glazing and the absorber plate is not all *that* critical. One

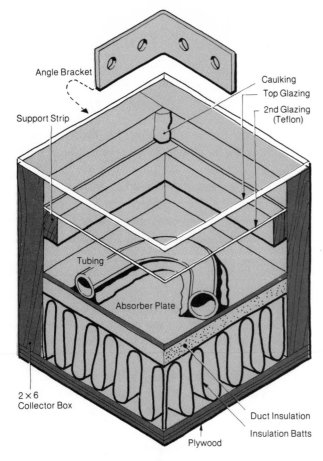

Figure 60. Here is an out-of-scale corner detail to show how everything in the collector panel might fit together. Notice how the corner is held together with an angle bracket and is caulked. Notice too how the fiberglass building insulation does not come in direct contact with the bottom of the absorber plate, but is protected by a layer of duct insulation. Resins in the fiberglass can vaporize in great heat and cloud the bottom of the glazing so less light can pass through.

inch is just about right, even though experts find very little difference between 1/2 inch and 1 inch. This airspace *does* need to be sealed off as well as possible to keep dust from coming in and settling on the plate. So *every* joint should be caulked. And when you buy caulking, don't choose a cheap run-of-the-mill type. Silicone caulking is better than average; so is "phenoseal," an acrylic latex caulking; "Butylcaulk" is good, too.

One of the things no one ever seems to mention is *venting* for the collector box. This is important. Make 3 or 4 small "weep" holes in the

lower side of the box so the box can breathe. Those needn't be larger than 1/16 to 1/8 inch in diameter. If there's a *slight* airflow to and from the box and between the layers of glazing, there will be less chance of condensation on the cover plate. Incidentally, there's no point in trying to make the glazing seal *perfectly* airtight because the glazing material — whatever it is — will need to expand and contract a little, constantly breaking the seal (Figure 61).

We already know that heat shouldn't be allowed to escape through the bottom of the box. This is why there must be insulation beneath the absorber plate. Three to 4 inches is plenty. *Don't* use ordinary batts of home insulation alone. Regular fiberglass insulation has a resin "binder" which vaporizes if it's heated to around 300 degrees F. The vapors from this resin will waft up past the edges of the absorber plate and condense on the underside of the glazing. There it will collect and look a little like maple syrup. This only blocks out light.

Put in a 1-inch layer of rigid fiberglass "duct" insulation — the kind without foil facing — immediately below the absorber plate. This looks like yellow fiberboard, has no such resin binder, and can be bought at any plumbing and heating supply house. Then use insulation batts *below* it

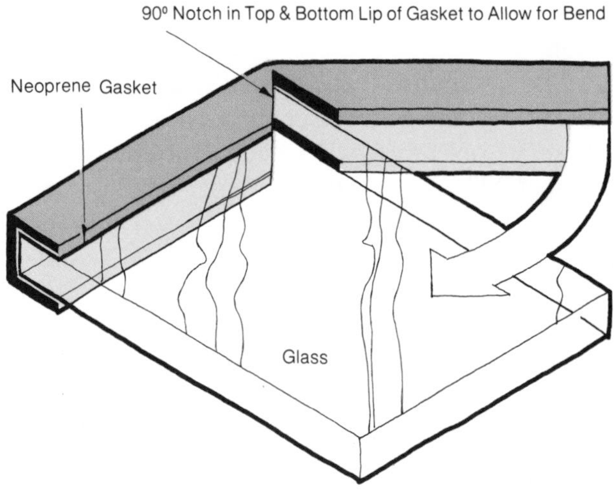

Figure 61. You might want to protect the edges of glass with a neoprene gasket. This should go all the way around the pane, and can be notched at the corners to give the appearance of a mitered joint. Gasketed glass has a superior seal, making the glazing waterproof and reasonably airtight.

if you want (see Figure 56). Foam insulation, by the way — styrofoam or urethane — will soften, melt, shrink or disintegrate if it gets as hot as 200 degrees F. Don't use it.

One more thought while we're on the subject: Wood is fair insulation itself. If you choose to build the collector box of *metal,* make sure the edge of the absorber plate doesn't touch the metal sides of the box. If it *does,* lots of heat will be conducted through the sides of the box to the ambient air. This means that any metal box should be insulated along its sides as well as underneath.

How Should You Select Glazing?

An absorber plate, no matter how good it is, can't make use of any radiation that doesn't get *to* it. So a glazing's ability to let light through is of number-one importance. To be really effective, a glazing material should be at least 88 to 92% translucent.

Most plastic glazings don't have the transmittance or heat-enduring properties of glass. In some cases, they get destroyed because ultraviolet rays break them down in a relatively short time. Others seem to hold up well enough, but certain ingredients in the plastic deflect rays sideways as they hit — rather than letting them straight through to the inside of the collector. Still, there *are* some notable exceptions, which make certain types of plastic even better than glass. Here are the four major types of glazing:

1. Hard Plastics. These are a good deal lighter than glass, and are easier to cut and handle. You're probably familiar with *plexiglass,* the most commonly known of the hard, transparent plastics. There are also vinyls. These are inexpensive, but they can only be used as *external* cover plates — and *only* when there is a second layer of different glazing material (glass or Teflon) beneath them. This is because they have what's called low "deformation temper-

atures" — a technician's way of saying that they'll bend and warp if they get too hot. Hard plastic is probably not a good bet for glazing in most cases (Figure 62).

2. Films. This category includes polyethylene, *Mylar, Tedlar,* and clear *Teflon* — all thin plastics. Polyethylene, although it's easy to get and cheap, can't take a lot of heat either, and doesn't transmit light as well as some of the others. At best it will last a year, and most people find it unattractive on a "permanent" solar collector.

Tedlar and *Mylar* are often mentioned in solar literature as fine glazing materials, but their worth is only marginal compared to glass and other things. They don't transmit all of the radiation they should to qualify as excellent solar glazing and they don't trap reradiated beams as well as glass. Mylar also degrades in the sun, though not nearly so fast as polyethylene. *Tedlar* has a life expectancy of about 6 years in most parts of the U.S., but it will gradually *shrink* in the heat, and shouldn't be stretched too tightly across the collector box for just that reason.

Teflon is excellent as an *interior* covering, but it only comes in a 1-mil thickness, making it too thin for an exterior cover "plate." (Any film should be at least 4 mils thick to be useful as an outer glazing.) Teflon is every bit as expensive as glass, but it's easier to work with. And it's practically invisible, letting through 98% of the light that hits it. It comes in rolls, but it can't normally be found in the average hardware store. The best source for clear Teflon at the moment is the DuPont Company in Wilmington, Delaware (see appendix). Order only as much as you need because it's quite expensive.

3. Glass. Glass comes in various strengths and grades. High-quality glass has a very low iron content, meaning that it reflects the least amount of light. Good glass should absorb only 3 to 4% of the radiation passing through it, whereas lower-quality glass may absorb 6% or more.

To judge the quality of a glass, look through it from its edge. The best glass will appear water clear or slightly bluish in hue. But poorer glass will have a greenish tinge. This may cost a good

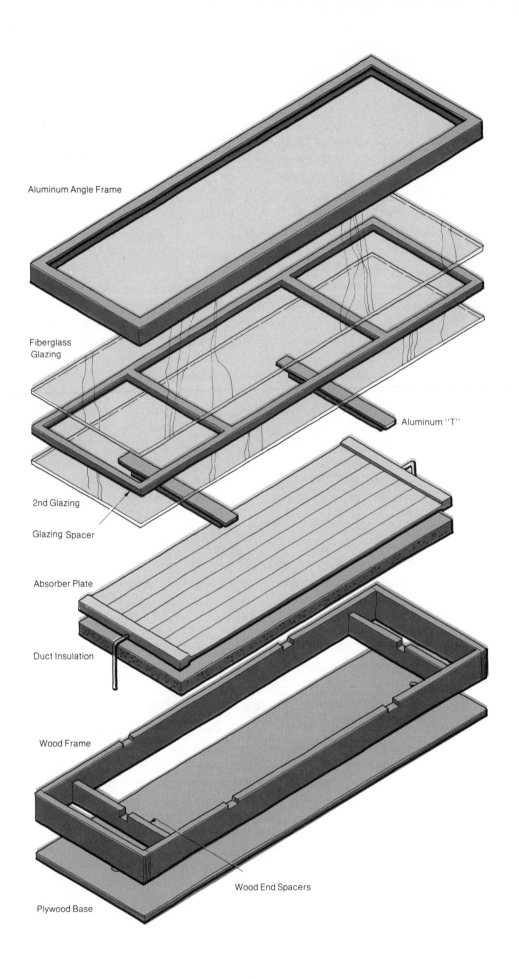

Aluminum Angle Frame

Fiberglass Glazing

Aluminum "T"

2nd Glazing

Glazing Spacer

Absorber Plate

Duct Insulation

Wood Frame

Wood End Spacers

Plywood Base

Figure 62 (opposite). This exploded drawing is taken from plans for a solar collector prepared by Garden Way. The top Kalwall fiberglass glazing is held in place on the wooden frame by a box made of aluminum angle. Note how the glazing is supported every 3 feet by 2 strips of aluminum T-bar. These must be notched into the sides of the wooden box. The absorber is the Kennecott Terra-Light copper collector plate, and it's supported by thick duct insulation. Study the detail to see how a second layer of glazing can be added, making the panel thicker. Detailed plans for complete solar hot water systems are available through Garden Way Publishing Company.

deal less, but it won't transmit light as well and is less suitable.

Most people choose to glaze solar collectors with single-strength glass, even though double-strength may be better. "Tempered" glass is super, but it must be ordered to size and, has to be treated with kid gloves. If you chip an edge, a crack will develop in time. If you want to spend the money, you can even buy low-iron glass that is treated with an antireflective coating. A.S.G. Industries in Kingsport, Tennessee is a good source for special solar glazing (see appendix).

4. Reinforced Polyesters. Most of us know these as *fiberglass,* even though the glass fibers are added mainly as strengtheners. *Glassteel,* a decent glazing, is a clear fiberglass laminated with Tedlar. *Filon,* manufactured in California, is that corrugated material you sometimes see serving as opaque skylight panels in barns, sheds and awnings. It transmits light exceptionally well, can be bought in flat rather than corrugated form, and is easy to find locally in many places.

Reinforced polyesters are as good as glass in many ways — in some ways better. You can't see through them, but they have many of the same properties as glass. They're ideal for the do-it-yourselfer because they don't crack or break under normal conditions, can be cut with a regular power saw, drilled, and even nailed. They also can't be shattered by hailstones or rocks.

A fine "fiberglass" glazing is *Kalwall Sun-lite Regular* or *Sun-lite Premium,* made in Manchester, New Hampshire (see appendix). It's 88 to 92% transparent to solar radiation and is recommended in areas where winter temperatures av-

erage 30 degrees F. or below. Two layers of Kalwall *can* be used, but normally it's limited to the outer glazing where there's a film such as Teflon underneath. (The Kalwall company by the way, has excellent customer service.)

In fact, all of the last 3 types of material can be used in various combinations if you choose to use a double glazing. You'd find very little difference in efficiency between them, but the very best combination would be a layer of low-iron glass above a layer of clear Teflon. Least efficient would be 2 layers of reinforced polyester. Somewhere in between would be a layer of fiberglass with a layer of Teflon, or 2 layers of glass. Take your choice. To be honest, all of these combinations are actually very close in efficiency, and all should work out fine.

How Should You Glaze Your Collector?

One layer of glazing or two? That's the great debate. There are plenty of arguments either way, and plenty of graphs to "prove" that one choice is better than the other. Most of these conflicting studies take all the factors — solar angle, climate, orientation, tilt, latitude and ambient temperature — into account. One thing, though, is pretty clear: If the temperature where you live is likely to fall below 0 degrees Fahrenheit at any time during the year, you'd better use a double glazing. The second pane will surely help reduce heat losses to passing winter winds.

It would seem that if two layers are better than one, 3 layers would be better still. Wrong. One layer of good glazing transmits about 92% of the radiation that hits it to the collector plate, but 2 layers only lets through about 80%. After that, the law of diminishing returns takes over again. Each layer of glass or plastic absorbs some heat, shades the collector some, and reflects some radiation back to the atmosphere. A double glazing, with an air sandwich between the layers, acts like a thermalpane window to hold in heat and keep cold out. But 3 layers, in spite of what some claim, is likely to be *less* efficient than 2 — by a lot. This is mainly because of the large reflection and absorption losses.

The primary reason for glazing, remember, is to prevent loss of heat energy through *convection*. Air movement pulls heat away from the collector plate in this case. The more airtight you can make the panel by sealing around the glass shield, the less air circulation there will be inside. But sealing and supporting the glazing are the most difficult problems.

The external layer is usually a rigid material — either glass or reinforced polyester. This can simply be laid on top of the 2 × 6 collector box. It should be fastened with some sort of gasket arrangement or caulked with an elastic caulking to prevent leaking and still allow for expansion (Figure 63). Caulking is easy enough to apply. It comes in a tube that fits into a caulking gun. With a little practice you can learn to squeeze the trigger of the gun to run a nice even "bead" of sealant to any surface the edge of the glass will be resting against.

Unfortunately, you can't have a long expanse of unsupported glass. Glass cracks or shatters under a lot of stress, but it's actually a very *elastic* material that will sag under its own weight if it's too long or too wide. In other words, don't try to cover a large collector with a single sheet of glass.

The maximum size you can use is about 24 inches by 72 inches, and that's pushing things. Aluminum "H" moldings offer one of the best

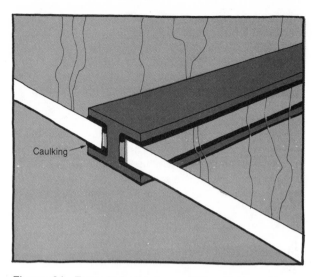

Figure 64. Extruded aluminum "H" molding offers a simple way to join glass edge to edge. A bead of phenoseal caulking run in the molding channel will make a fairly good seal. Don't forget to allow room for the glass to expand.

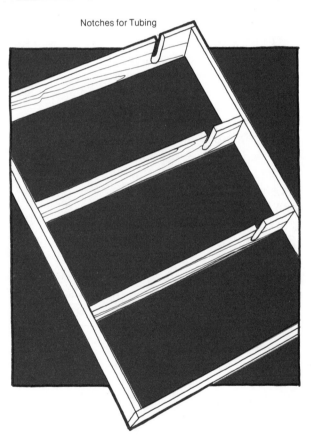

Figure 65. When there are several collectors to be mounted together, make a single framework divided into separate bays for the individual absorbers. The dividers will have to be drilled or notched for the tubing that connects the panels.

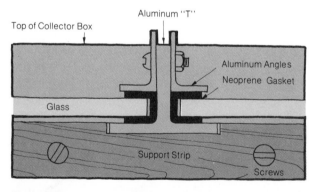

Figure 63. Glass that has been gasgeted with neoprene can be supported and joined with this kind of arrangement using aluminum angle. This way the glass will be slightly recessed in the collector box.

ways to join panes of glass edge to edge (Figure 64). If you're going to have many solar collectors together, it might be wise to build one large collector box with dividers to house the individual absorber plates. This way you can glaze with larger sheets of fiberglass and shiplap your joints, sealing them with caulking (Figure 65).

The inner layer of glazing can be either glass or plastic film. The film should be stretched between the sides of the collector box, pulled over a strip of wood and stapled before the wooden strip is nailed to the inner side of the box. If the inner layer is glass, the edge of the glazing might rest on wooden support strips, or be recessed into the sides of the box. The bite of the slot should be as deep as it can be, to support the glass as well as possible and still allow for expansion. Don't forget the inner glass will expand more than the outer layer because it's hotter in there.

What About Spacing?

There should be about 1/2 inch of space between layers of glazing and at least 3/4 of an inch to 1 inch of space between the inner glazing and the absorber plate.

If you do choose glass as your outer collector covering, install it last — *after* the rest of the collector assembly has been fastened to the roof and all the plumbing is in place. Then get some of those suction cup devices with handles on them that glass companies use when they carry plate glass windows. Using them is the best way to lift glass onto a roof. They not only make the glass easier and safer to handle, they keep it from getting smudged with fingerprints. Obviously you want the glass to be as clean as possible once you get it on. And try not to let any foreign material into the collector box while you're working. Once the glass is in place, you should have nothing to worry about except covering the collectors until the rest of the plumbing is finished.

When the whole system is in operation, you might want to go up on the roof once a year or so to admire your handywork. While you're there, squeegee off the glass to remove any oils that might have collected out of the polluted air. Dust may gather on the outer glass from time to time, too, but it won't slow the collector down noticeably. Rain will take care of it eventually, anyway.

CHAPTER 6

Storing
and Protecting
Your Solar Energy

We know how to collect energy from the sun. Our problem now is to *store* it effectively. Solar energy, as everyone knows, is not constant. We have seasonal changes and clouds, which naturally affect the sun's intensity, and we also have night. One reason why heating domestic hot water with the sun makes so much sense is because water itself is such an excellent storage medium. In fact, if you're going to transport or store solar energy, using water — or at least a fluid containing water — is one of the best ways to do it.

Earth, rocks and concrete are also fine storage mediums. But they're not as good as water — nor nearly so versatile. You see, every substance, whether it's a solid, liquid or gas, has a characteristic called its *specific heat.* Its specific heat is the number of BTU's needed to change the temperature of that substance 1 degree Fahrenheit.

Water has a specific heat of "1," meaning that it takes 1 BTU to raise the temperature of 1 pound of water 1 degree F. The specific heats of all other substances are measured against water. In fact, the whole concept of the British Thermal Unit (BTU) is based on the idea that water is the standard against which other things can be compared.

Water has a very stable disposition as far as temperature change is concerned. You can feel this if you wrap your hand around a glass of ice water or a cup of hot coffee. The warmth in your hand does very little to heat the cold water in the glass, and there is no way you could significantly reduce the temperature of the coffee — at, say, 180 degrees — to 98.6 (your temperature) just by keeping your hand around the cup. Water is so common we take it for granted, but it's unique stuff because it *is* so stable.

A non-toxic antifreeze like propylene glycol — which we'll need to discuss soon — has a specific heat of 0.6 BTU's per pound. Rocks, iron, concrete and many other solids need only about 0.2 BTU's to raise their temperature 1 degree. So their specific heats are 0.2. It takes more BTU's to heat water (1.0 as compared to 0.2), but by the same token it takes more to cool it off (also 1). This is just another way of saying that more heat can be stored in a substance with a high specific heat than can be stored in one with a low specific heat.

Even in this modern day and age, with all our technology, an insulated tank full of hot water is still one of the best ways to preserve many BTU's of energy. A single cubic foot of water can store 62.5 BTU's of heat per degree Fahrenheit. Stone, by comparison, has a heat-holding capacity of 36 BTU's per cubic foot per degree F. — a little over half that of water. But air — sometimes used as a medium to transfer solar energy — only holds 0.018 BTU's per degree F. per cubic foot!

What Should Your Storage Tank Do Besides Store Heat?

Let's discuss some terms you'll want to be familiar with as you design your solar heater:

People who work with solar heating say that a system is *down* whenever the sun is not shining and all the stored solar energy has been used up. In other words, "down" means the system is just not producing. The *down point* is reached when hot water can't be taken from the storage tank any longer — because it's not hot enough. At this point we either go without a hot shower or have a set-up that allows a fuel-firedor electrical back-up system to kick in.

Thermal lag describes the amount of time it takes to collect enough heat — once the sun is shining again — to get water in the storage unit above the down point. (The "lag time" is how long it takes to make the system productive again.) The term *carry-through* describes the amount of time *without* sunshine that a solar heating system can continue to function without help from any other heat source. A system with a long carry-through is probably very well insulated with an oversize storage capacity, as well as having efficient collectors to feed the storage. Obviously we want any system to operate above its down point as much of the time as possible, as well as to have a brief thermal lag.

You'll need a larger storage tank for a solar water heater than you probably now have for your conventional set up. It's smart to assume there will be fairly long periods when there's little or no solar energy coming in. The larger the tank, the less likely you're going to run out of solar hot water.

Allow enough storage capacity to keep each member of the household in hot water for at least 2 sunless days. (That sounds like a lot, but it's only 40 gallons a person, if you figure 20 gallons per person per day.) Two people can get away with an 80 gallon tank. But any more than 2 will need a 120 gallon tank. If you can afford it and have the space, always buy a larger tank than you think you'll need.

If you live in a northern climate, be realistic. Think of your solar water heating system as a "preheater" for your regular hot water system, even though you'll be trying to provide enough hot water from the sun to keep the electric, oil or gas hot water heater from working most of the time. The two systems, old and new, should be connected — or "in line" as they say. As the solar-heated water leaves its own storage tank, it should pass through the conventional water heater, but be so hot that the thermostat in the heater registers no need for the heat to come on. But face it. There will be times when your back-up system *must* come on, even if you have a huge storage capacity (Figure 66).

It would be a mistake to buy a very large storage tank with an electrical heating unit — called a "booster" — right in it. Because electricity will heat water so much faster than it can be heated in the solar panels, the electrical element will keep coming on before the collectors have had a

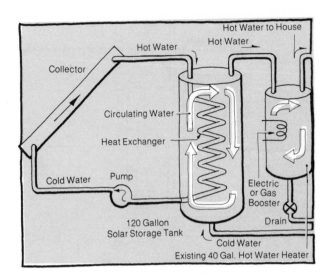

Figure 66. Think of your solar water heater as a pre-heater for your existing hot water system. The two systems should be connected in a series so that hot water flows through both tanks on its way to the house. Don't forget that the existing hot water tank will also store solar hot water. This is why it too should be insulated every bit as well as the primary storage tank. Water in both tanks will be circulating much of the time. As the hottest water comes in contact with the walls of the tank, some of its heat can be radiated to the outside air. This is why there must be insulation all around the tanks.

chance to restore enough heat to the tank. In other words, the electric heater will *interfere* with the solar system, and throw into a tizzy any thermostatic controls you may have. It's far better to have two tanks: one exclusively for the solar system, and one for the back-up system, even though heated water will pass through both.

It's quite possible that your system could be "down" if you washed a lot of lunch dishes and then a couple of your house guests decided to take long showers. The thermal lag — or recovery time — would then depend on the capacity of your collectors. On a good bright day it's reasonable to expect that the collectors can *restore* hot water in the tank at a rate of 1 gallon per square foot of collector surface per hour under ideal noontime conditions. So if you had two 24 square foot collectors, you might anticipate up to 48 gallons of new hot water in an hour's time.

Warning: Water in solar storage can get *too* hot if it's sunny and you don't draw any water from the tank for several hours. Kids can scald

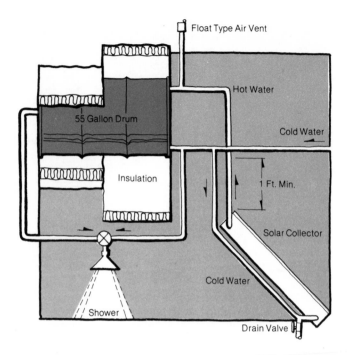

Float Type Air Vent

Hot Water

Cold Water

55 Gallon Drum

Insulation

1 Ft. Min.

Solar Collector

Cold Water

Shower

Drain Valve

Figure 67. Here is a schematic diagram of a simple solar shower for a cabin or summer camp. No anti-freeze is used, and there's no pump. Because this is a thermosyphoning system (see Chapter 8), the bottom of the storage tank — in this case an insulated 55-gallon oil drum — must be at least a foot higher than the top of the solar collector. There must be a drain valve at the lowest point in the system so the plumbing can be emptied completely when it's not in use. Notice, too, that the vent pipe must be placed at the highest point in the system. If it's not, water will flow out of it.

their hands if there's no water temperature regulator. Teach them to mix water at the tap when they wash. Turn on the cold first, and *then* the hot — gradually (Figure 67).

How Should You Select a Tank?

Spend some time shopping for your storage tank. They vary a lot in price, quality, and in the amount of insulation they have. Used tanks of various kinds may offer some attractive possibilities. Look for old water tanks or larger propane tanks if you can only spend a little money.

The tank will need to stand a lot of pressure — as much as 60 pounds per square inch (psi). Sixty psi is normal in many homes and municipal systems. So check the tank's structure pretty thoroughly, particularly joints or welds you make yourself.

The outside surface area of the tank is something to consider too. A spherical tank would be ideal because it has the least amount of surface, but nobody makes them so far as we know. A short squat cylinder has less surface area than a tall, skinny one. Ideally you should get the diameter of the tank as close to the height as possible.

In the case of hot water storage, convection becomes your worst enemy. It's a real heat thief. Because the hottest water in the tank always wants to rise, and the coldest water in the tank wants to settle to the bottom, the contents of the tank keep circulating as new water of differing temperatures passes in and out. As the water circulates, the hottest water is continually exposed to the outside of the tank where its heat can be lost to the outside air (See Figure 66). If there's less surface area, of course, less water is exposed to the sides of the tank.

Heat has a natural tendency to *radiate* from a hot place to a cold place. As a matter of fact, this is one of heat's major characteristics, regardless of where it's found. It will *always* travel from an area with the highest temperature to an area with the lowest temperature. And heat loss is "omnidirectional," meaning that it will escape in any direction it can.

Stable as water is, heat wants to drift away from *it,* too. The purpose of wrapping extra insulation around the outside of a hot water tank, of course, is to block the heat's path of escape. It's also important to know that some radiated heat can be reflected *back* at its source by using foil-faced insulation (as long as the foil faces in — toward the tank).

If you live in a temperate climate where the temperature may fall below freezing during part of the year, you must circulate a mixture of water and non-toxic antifreeze through your collectors so the panels can never freeze. This mixture has to be kept separate from your regular water supply, naturally.

This means that your system will have a loop of plumbing to run antifreeze fluid through the col-

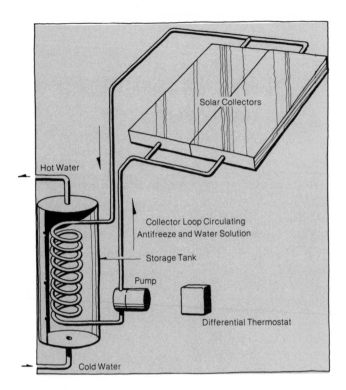

Figure 68. In climates where there is even a remote chance of freezing, or wherever there's very hard water, which can leave mineral deposits in the system, a fluid containing antifreeze and water must be circulated through the solar collectors. Heat is given off from the collector loop through a coil of tubing called a *heat exchanger*. Although the heat exchanger may be right inside the solar storage tank, the two fluids never actually come in contact with each other.

Look for a tank-heat exchanger combination that's stone lined. The stone inhibits corrosion of the metal tank walls, even though it won't necessarily help the tank hold heat more efficiently. A good one should last many years, and some even have a 10-year warranty. Glass-lined tanks, incidentally, don't last as long as they should for this kind of a system (Figure 69).

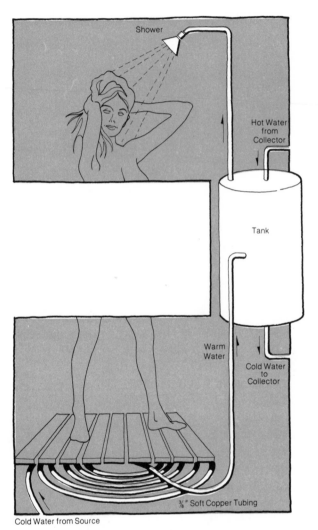

Figure 69. Heat exchangers can be located in chimneys, fireplaces and woodstoves so that water passing through them can pick up additional free heat. In this case, a coil is placed in the floor of the shower. The solar heated water from the shower head does double duty. It not only warms and washes the person in the shower, it contributes some of its heat to the cold water on its way to the storage tank.

lectors, and a second layout of plumbing to run pure water from the supply line to the storage tank, through the back-up system and to the house itself. The first loop is usually called the *collector loop*; the second is called the *hot water loop*. Heat must be passed from one loop to the other at what's called a *heat exchanger* — a coil of tubing that's usually within the storage tank itself (Figure 68).

Heat exchangers will be explained in detail later on, but for now let's just say that a large storage tank with a heat exchanger coil — or a *second* heat exchanging tank inside the larger outer one — are fairly easy items to find these days (see appendix). It's probably not a good idea, by the way, to buy an old tank and try to build in a heat exchanger coil yourself. Nearly always this turns out to be more expensive than buying a new one.

Ford Products of Valley Cottage, New York (see appendix) makes fine stone-lined heat exchanger tanks that are guaranteed for 5 years and should last for 20. Their 120-gallon tank (Model TC-120) is just about ideal for a family of 3 or more — in both price and storage capacity.

Your solar water tank should be insulated as perfectly as possible. But don't forget that the conventional water heater is also holding hot water. *It* may be open to tremendous heat loss. Don't assume that your "insulated" electric water heater, for example, is insulated enough. It probably has only 1-1/2 inches of fiberglass insulation. You will be wise to give it another 3-1/2 inch wrapping of foil-faced or kraft-faced insulation held in place with duct tape and staples.

How Should You Install the Storage Tank?

Water tanks are heavy. The water alone in a 120-gallon tank weighs about half a ton. (A cubic foot of water weighs 62.5 pounds, and there are just a touch over 7.5 gallons to the cubic foot.) Naturally it would be dangerous to put a large tank in an attic or on some wooden floor that might not support it. This is why most solar storage tanks wind up either in a basement or in a heated garage sitting on a concrete slab.

Ideally the tank should be situated slightly away from walls and corners, so you can walk all the way around it. This will make repairs and installation of the rest of the system a whole lot simpler. If you don't have enough room to do this, plan things so that important valves, switches, gauges or connections are out front and easy to get at.

At some point — we hope in the far-distant future — the storage tank is going to corrode through somewhere and spring an irreparable leak. (They all do sooner or later.) Don't forget to plan an escape route for the dead tank. It would be embarrassing to have to jackhammer out a basement wall just to get it out and get a new tank in.

How Should You Insulate Your Storage Tank?

Keeping heat in a hot water storage tank is like trying to keep a leaky inner tube pumped up. No matter how hard you try, you're constantly fighting a losing battle. Some heat will always escape.

How much insulation is enough? It's theoretically *impossible* to over-insulate a solar hot water system, but here are some minimums: There should be *at least* 3-1/2 inches of fiberglass insulation — or its equivalent — behind the absorber plate in the solar collector, as we already know. There should be at least 1-1/2 inches of insulation on all pipes in the system, and there should be at least 6 inches of insulation around the storage tank.

If you can afford to do more than this, so much the better. Protecting your solar-heated water with more-than-adequate insulation (that has been carefully installed) will reward you directly because your back-up system will come into play far less often. You should also have a basic understanding of how much heat leaks through various types of insulation. Here's how *those* things are computed:

Every material has a measurable "C" value and an "R" value. The "C" value describes the *conductivity* of an insulation. It tells us how many BTU's of heat will pass through a square foot of the material in an hour's time when the temperature on one side of the material is 1 degree F. colder than on the other. You have no convenient way to measure C value, and you won't know what it is because it's rarely printed on the label of insulation packaging.

But the "R" value — which *will* appear on the package — is something you can know about. The R value is dependent on the C value. R tells you the *resistance* of the insulating material to heat passing through it. In mathematical terms, R is the reciprocal of C. In other words,

$$C = 1/R$$

For example, if you read that the R value of a particular insulation is 20, that means that 1/20th (0.05) of a BTU of heat will pass through 1 square foot of the material in 1 hour if the temperature differential is 1 degree Fahrenheit. If the package says R-11, that means that 1/11th (0.09) of a BTU will pass through in the same amount of time, and so on.

To put it another way, the greater the R value, the greater resistance the material has to the passage of heat. So an insulation layer with a high number — 20 for instance — is better protection than one with a low number like 4. Unfortunately, doubling or tripling the thickness at a material doesn't necessarily produce twice or three times the R factor.

R value usually accounts for an inch thickness of whatever material we're talking about. One inch of fiberglass building insulation has an R value of 4. But 3-1/2 inches of the same material has a value of only R-11. Six inches jumps to R-19, and 12 inches all the way to R-38. Styrofoam is rated at R-4.2 per inch — a little better than fiberglass. But urethane foam is better — R-6.3 per inch. Wood (pine specifically) has an R value of only 1.2, but concrete, metal and stone are much worse insulators — R-0.08 (Figure 70).

There's a third value to keep in the back of your mind. The "U" value is the *total* of all the heat resistances of all the materials surrounding the hot water — or any other heat source that's being insulated, for that matter. This includes the tank's metal sides, the insulation and the airspace if there is any. Again, U is a reciprocal of all the R's,

$$U = 1/R$$

Say the total R factor for one square foot of insulated tank comes to 20.2 (including the metal, airspace, 6 inches of fiberglass and some wood). The U value will be 1/20.2 or 0.049. (This means, to say it one more time, that 0.049 BTU's per square foot can escape from the tank when the temperature differential is 1 degree F.)

If the water in the tank is 140 degrees and the temperature in your basement is, say, 60 degrees, the temperature differential is 80 degrees (140 − 60 = 80). Multiplying the U value times

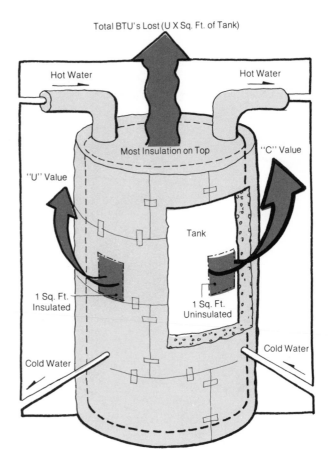

Figure 70. Heat loss from a storage tank is continuous, and it can happen in any direction. To insulate a tank as effectively as possible it's good to be familiar with the *conductivity* (C value) and the *resistance* (R value) of the various materials surrounding the water. To calculate total heat loss, you should also know the U value, which is the total of all the resistance in the insulating materials.

the temperature difference (0.049 × 80) we get a heat loss per square foot of 3.9 BTU's per hour.

If the total square footage of a 120-gallon hot water tank is around 50 square feet, then the total heat loss from the tank is 50 times 3.9 BTU's or 195 BTU's per hour. This is a very round-about way of saying what was said originally, that no matter how well you insulate your tank, there is still going to be a certain amount of stored solar energy lost.

Because heat tends to rise, insulation does the most good above the heat source. Any extra material should be put at the top of the tank. Another way to insulate a tank very well is to build a

65

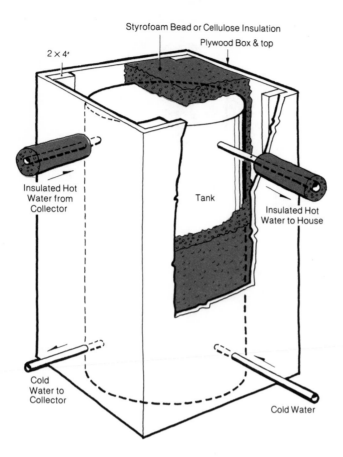

Styrofoam Bead or Cellulose Insulation

Plywood Box & top

2 × 4'

Insulated Hot
Water from
Collector

Tank

Insulated Hot
Water to House

Cold
Water to
Collector

Cold Water

Figure 71. One of the best ways to insulate a solar hot water tank is to build a plywood insulation bin around it. This can be filled with styrofoam beads, sheets of urethane foam, blown cellulose insulation, or foamed-in-place insulation.

plywood box around the tank and fill it with styrofoam beads (R-3.6 per inch) or the kind of cellulose insulation that gets blown into the walls of old houses. Blown cellulose has an R value of 3.7 (Figure 71).

Foamed-in-place insulation (R-5.2 per inch) is better still. It fills all the nooks and crannies around the tank, and it's just a matter of mixing two liquids together and pouring them into the box. The mixture will expand, foam and bubble like a witch's brew, and work its way into all the spaces between the tank and the plywood. A good source for this stuff is your local insulation contractor.

These last two methods although they're quite good, have certain disadvantages too — not the least of which is that the tank is awfully hard to get at once the insulation is in place. In any case, take some of these thoughts and wage your own war with heat loss as best you can.

66

CHAPTER 7

Building the
Transfer System

The energy transfer system in a solar water heater is nothing more than good old fashioned plumbing. We move heat energy from the source (the absorber plate in the collector) to the storage area (the hot water tank) through pipes. This transfer system may be regulated in some cases by a control system, which usually takes the form of a pump, some valves and a differential thermostat, but this is not nearly so complicated as it sounds. (All of these last things will be discussed in Chapter 9.)

There is nothing fancy or exotic about the plumbing layout. If you've had any experience at all with measuring, cutting pipe, and soldering joints, you should be able to do all or most of the work yourself.

Needless to say, you should plan as much as you can in advance. Remember the advice of the Wright Brothers, who said in effect, "If you get it right on paper before you start, it'll be right when it's built."

If you're totally inexperienced, you can always consult a licensed plumber whenever your confidence falters. For most plumbers, solar water heating is still something of a novelty. If you call one you know, he may be interested to learn about what you're doing. Who knows, you may even get some free advice — as we have from time to time.

You may be very tempted to use plastic tubing because it produces very little in-line friction and because it's so easy to cut and install (also because some other sources will *recommend* that you use polyvinyl chloride — PVC — or chlorinated polyvinyl chloride plumbing that you can glue together). *Don't.*

Plastic plumbing can't stand the heat in a high-performance solar water heater (although we *do* recommend plastic tubing for a solar system to heat swimming pool water — see Chapter 10). Most of the joints and fittings will be fine until one blistering hot day when there's lots of sun. When the water temperature gets very high — say 200 degrees — something will give out, and suddenly you'll have a serious leak and a major repair job. Sometime soon there will be plastic plumbing systems that *can* withstand these temperatures, but they're not here yet. Stick with copper for now.

Copper tubing comes in two basic forms. Normal household plumbing is usually made up of the first — *rigid copper tube.* You might want to use this if you're putting a solar hot water system in a new house. It's straight, easy to work with and more attractive than the second type — if it's exposed to view. Rigid copper tubing is normally sold in 10- or 20-foot lengths.

Soft copper tubing is sold in coils, 15, 30 or 60 feet long. Use this if you're planning to retrofit an old house. It's much easier to thread through old walls and other tight places because it's flexible. And because it comes in such long pieces, you don't have to worry about making a joint in some hard-to-get-at spot.

It's a good idea to be sure of the *grade* of tubing you're buying. The outside diameter (O.D.) is always 1/8 inch larger than the stated diameter. That is, 1-inch pipe is actually 1-1/8 inches in O.D. But the *inside* diameter varies, depending on the grade (Table 10).

Type K has the thickest wall. It may last longer, but it will transport less fluid because the inside is smaller. The price is also exorbitant. Type L has medium thickness. Type M has the thinnest tube wall, but it's very difficult to find in soft copper rolls. In a simple solar water heater with natural water circulation, soft Type L tubing is usually recommended because of the corrosion factor with plain water. But if you're planning a forced circulation system, rigid Type M, the lower grade tubing, should be perfectly adequate because you'll be using a low corrosion heat transfer fluid — the antifreeze solution.

The *flow rate* of liquid through a pipe is expressed in gallons per minute (gpm). Flow rate, of course, is reduced when there's lots of friction within the tube. (Plumbers use other terms to describe friction, by the way. They call it "pressure drop" or "head loss.")

Normal home plumbing rarely uses tubing that's less than 3/4 inch. That should be just about right for most forced circulation solar water heaters — although for a very long run or for a thermosyphoning system (see chapter 8) 1-inch pipe would be better. (Pressure drop in 100 feet of 3/4-inch copper pipe at a flow rate of 12 gallons per minute is 15 pounds per square inch. But in a 1-inch pipe, equally long, with water flowing through in at the same rate, the pressure drop is only 4.25 psi.) For a set up with a very large capacity you may want 1-1/2- or even 2-inch mains.

Another important lesson in the general-plumbing-information category: A lower flow rate reduces friction and thereby reduces pressure loss. So you'll want to find a happy medium. A typical flow rate through a solar water heating system is at the most 5 gallons per minute — relatively slow. If it's faster than one gallon per minute per collector, you're endangering the collector by exposing the fluid channels to something called "erosion corrosion." By the same token, each collector should have a *minimum* flow rate of 1/2 gallon per minute.

How Should You Lay Out Your Plumbing?

Again, sketch everything on paper before you make a move, reminding yourself that the run of plumbing between the collectors and the tank should be as short as possible. There are three very good reasons for this: (1) you want to eliminate as much heat loss as you can, (2) you want to reduce friction in the pipes, and (3) you want to save the cost of tubing.

The shortest distance between two points, as always, is a straight line. This goes for plumbing too. Try to use as few elbows and 45-degree angles as you can. Friction in the pipes (head loss) will always be less if you make gradual direction changes instead of sharp ones (Figure 72).

And, don't forget this basic rule of plumbing too: Put a boiler drain at the very lowest point in the system. Make everything flow downhill toward it, and don't leave any places where the tubing is level enough to "hide" water after the system has

TABLE 10

NOMINAL DIMENSIONS FOR TYPE L COPPER TUBING

Nominal Size	Outside Diameter	Inside Diameter
¼"	0.375"	0.315"
⅜"	0.500"	0.430"
½"	0.625"	0.545"
⅝"	0.750"	0.666"
¾"	0.875"	0.785"
1"	1.125"	1.025"
1¼"	1.375"	1.265"
1½"	1.625"	1.505"
2"	2.125"	1.985"

Courtesy George Daniels, *Home Guide to Plumbing, Heating and Air Conditioning.*

Table 10. This table presents the actual dimensions of Type L — medium weight — copper tubing. Notice that the outside diameter is always ⅛ inch (0.125) larger than the stated diameter. This is often a source of confusion.

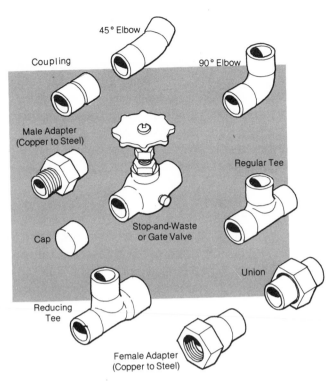

Figure 72. These are the most common sweat-type fittings used with copper tubing. The illustration pretty much explains the uses of each.

been emptied. Standing water can freeze, burst a pipe or blow a joint apart.

By the same token, it's important not to have level spots or sags in the lines where air bubbles can be trapped. When a tube gets "air locked" in this way, the water flow is slowed down or, in some cases, stopped altogether. If this happens in or near a collector, you can expect serious damage from overheating.

The very *highest* point in the system — somewhere in the out-flow header above the topmost collector — should have a float-type air vent. This will allow air bubbles to escape and prevent water from being syphoned out of the collector too rapidly. It also allows air to come into the system when you're draining it. With no vent it would be like trying to let lemonade out of a drinking straw when you're holding your finger over the top. Liquid won't flow downward in a tube unless air can enter at the top.

Your plumbing system also may need some sort of pressure relief. There will be times when you won't use hot water on a sunny day, and the fluid circulation will be very slow as a result. At times like this the fluid in the collector loop will expand, building up enough pressure to destroy a collector or some other part of the system. An *expansion tank* tapped into one of the lines will take up this overflow. In fact, the expansion tank's function is to keep consistent pressure in the system at all times — normally about 15 psi.

Don't forget that domestic hot water lines and collector-loop lines, which contain non-toxic antifreeze, must be kept totally separate. As you make connections close to the water tank, the two lines may be near each other. You might want to color-code each circuit so there is no possibility of confusion at some later date.

Working in an old house presents a problem for any new plumbing. Rather than try to run your tubing through existing walls, why not make hollow "posts" of 1 × 6 or 1 × 8 pine, which stand against a wall. These don't look bad if they're done well. They can be stuffed with fiberglass insulation and painted or papered to match the wall behind (Figure 73).

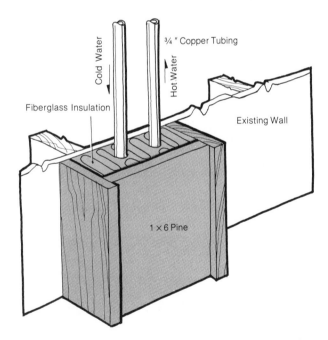

Figure 73. There's no reason for solar plumbing to be exposed, even in an old house that's being retrofitted. Soft copper tubing can sometimes be threaded through existing walls. But where it can't be, make hollow wooden "posts" to disguise the pipes. These should be stuffed with insulation to prevent heat loss. It's a good idea generally, to run the feeder line to the collectors and the return line close together. It saves work in the long run.

And while we're on the subject of insulation, we should emphasize that every *inch* of plumbing, if possible, should be protected against heat loss. One of the easiest — and most efficient — ways to do this is to run your feeder line and your return line parallel and close together — about 2 inches apart. Then wrap some ordinary fiberglass insulation around both pipes and staple it to itself.

A more expensive but more pleasant-looking alternative would be to buy flexible foam pipe insulation. *Vascocel* is an excellent product which you can buy at almost any air conditioning or refrigeration outlet. *Arma-Flex* is another very good name (see appendix).

Ready-made pipe insulation like this usually comes in 4-foot lengths and is slit end to end. You pull the insulation apart with your fingers, slip it over the pipe and stick it back together by wrapping it with waterproof tape. It's made to fit standard copper tubing exactly, so make sure you get the right size.

Finally, whenever piping must pass through the roof or exterior walls, make "X" slits in the insulation instead of holes, push the pipe through, and then gently stuff the insulation back around the pipe. This should give you the least amount of heat loss at that point.

How Do You Put the Plumbing Together?

Connections between pieces of copper tubing are made with special fittings and solder, as everyone knows. Plumbers, fix-it men and do-it-yourselfers call it "sweating" joints. Copper pipe, of course, is soft and easy to cut. You can use either a hacksaw or a small tube cutter which leaves you with a perfectly square end.

To sweat a joint you'll need four things: (1) some steel wool or emery paper, (2) flux, which is a paste that looks like Vaseline, (3) solder wire, and (4) a heat source — usually a propane torch. These things can be found at any hardware store. It's best to use regular 95-5 lead-tin solder (the type without a flux core), and acid-type solder.

Making a good leakproof joint is easy. Clean the end of the tube to be connected and the inside of the fitting thoroughly with the steel wool or emery paper. You just have to get the surfaces shiny. Don't sand them down too much or you'll destroy the tight fit. This should be done *immediately* before you smear on the flux. (If you wait even as little time as an hour before soldering, the surfaces will have time to oxidize again, and you may not get a true bond.)

Use your finger to apply a generous film of flux to both surfaces, and fit the joint together. Now heat both the end of the tube *and* the fitting, while you touch the end of the solder wire to the joint. (Try not to get the flux so hot that it starts to sizzle.) Be sure to heat the connection thoroughly on both sides.

When the temperature of the copper is right, the solder will suddenly melt and flow evenly into and around all sides of the joint. If you're doing this for the first time it may seem like magic to you, since the solder even flows uphill as a result of the strong capillary action created by the flux. If everything has gone well, you should have a perfectly watertight joint (Figure 74).

Don't be stingy with solder. It's cheap, and if you drip a little on the floor you won't have lost that much. A drippy joint, where too much solder has been used, is better than one where there's too little — even though it may not look quite so professional. With practice you can learn to wipe the joint quickly with a rag before the solder has set up. But be careful at first not to burn yourself.

One of the trickiest places to make plumbing connections is immediately under the roof. Here you have to work with the torch close to the rafters and roofing — which can be a fire hazard. It might be a good plan to stick a piece of asbestos or aluminum flashing temporarily between the joint and the wood so you don't burn the house down with the propane torch.

Automobile radiator hose and hose clamps, by the way, can get you out of some tight plumbing jams. These might be useful when you're making those connections through the roof to your header pipes (Figure 75).

Once everything is together, you should test your system for leaks before you run any fluid through it. It's virtually impossible to solder or resolder lines that are filled with water. The water

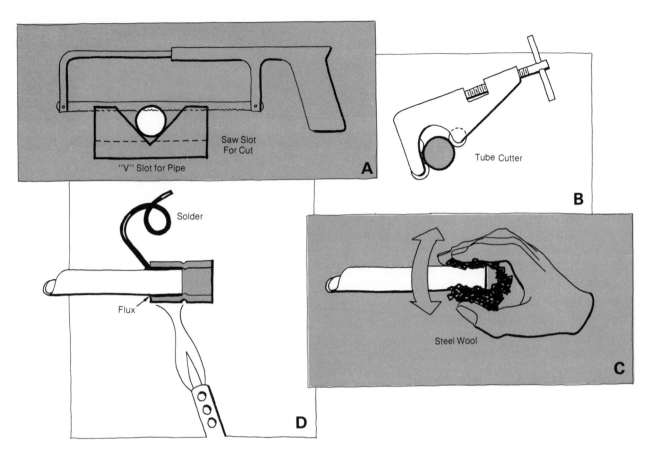

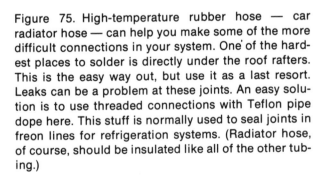

Figure 74. Copper tubing *can* be cut with a hacksaw, but you'll do a neater job if you set the tube in a pre-made "jig" first (a). A small tube cutter (b) is a fairly inexpensive tool, and is easy to use. It makes a perfect cut every time. Before a connection is made the outside of the tubing and the inside of the fitting have to be cleaned with steel wool or some fine abrasive paper. This is done best with a rotary motion (c). After the flux is applied, the joint should be fitted together and heated. Don't try to heat the solder wire directly. It will melt into the joint whenever the copper has reached the right temperature.

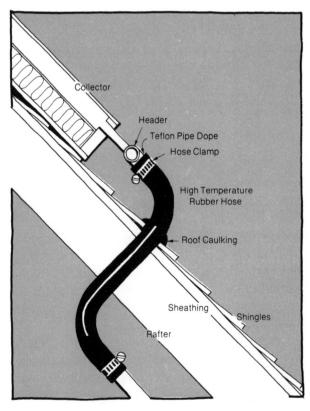

Figure 75. High-temperature rubber hose — car radiator hose — can help you make some of the more difficult connections in your system. One of the hardest places to solder is directly under the roof rafters. This is the easy way out, but use it as a last resort. Leaks can be a problem at these joints. An easy solution is to use threaded connections with Teflon pipe dope here. This stuff is normally used to seal joints in freon lines for refrigeration systems. (Radiator hose, of course, should be insulated like all of the other tubing.)

71

cools the surfaces too much to allow the solder to flow properly. Nothing is more annoying than having to drain a whole system and start all over again if you discover one leak.

The easiest way to test for leaks is to use air pressure. Thread a small "snifter valve" — a small fitting to which a hose can be attached — into a 1/8 inch tapping anywhere in the lines. (This can be done at any special tee that has a threaded hole to accept a gauge, thermometer or air vent fitting.) Attach the air pump hose to the snifter. Pump the lines full of air to a pressure of 20 psi, and then paint each joint with a mixture of soap and water. Check every connection for air bubbles, and resolder any one that looks suspicious. Once there are no air bubbles and the system has held 20 pounds of air pressure for an hour or so, everything is probably fine and you can remove the air pump.

How do you fill the system for the first time? Rent or borrow a centrifugal pump, jet pump or sump pump. Open all the valves in the lines — if you have them — and connect the pump hose at the boiler drain in the base of the system. Once fluid has been pumped in and circulated through the system at a pressure of 15 to 20 psi, all valves can be closed gradually, and the hoses for the centrifugal pump removed. From then on, your regular pump (or the thermosyphoning effect) can handle the circulation (Figure 76). (More on regular circulation pumps in Chapter 9.)

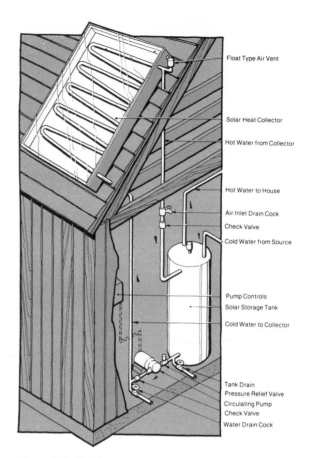

Figure 76. When you're all finished your system might look something like this. Notice the drain at the bottom of the system and the pressure relief at the top.

What About Antifreeze?

Don't use ordinary — and poisonous — automotive antifreeze. The very best antifreeze for solar collectors is something called "non-toxic propylene glycol." You should be able to find it in any plumbing supply store. If not, order *Sunsol 60* from Sunworks, Box 1004, New Haven, Connecticut 06508. (Also see appendix.)

Five gallons should be more than enough for a two to three-collector system in a normal house. (You'll need just about 1/2 gallon of unmixed propylene glycol for each 24-square-foot collec-

tor.) Other special mixes come already blended to protect your collectors to at least −34 degrees Fahrenheit.

Naturally the whole collector loop should be "flushed" every few years and the solution changed, just the way you would change the antifreeze in your car's radiator. Propylene glycol is only effective for 2 to 3 years. If you use some other solution — specially earmarked for solar systems — read the label on the container to learn how often it should be changed.

One last bit of advice: In spite of what the Wright Brothers said, plumbing *doesn't* always go exactly as planned — no matter how complete your sketches are. (We all know what the poet Robert Burns said about the "best laid plans of mice and men . . .") Be flexible. And if you get in a real snag, don't be too proud to ask an expert for help.

CHAPTER 8

Free-Circulating Systems and Heat Exchangers

Natural circulation solar hot water systems — *thermosyphon* systems, to use the more technical term — have a lot of appeal for households that don't need very much hot water. We've touched on these a bit already. The real beauty of a free-circulating system is that it operates with no fuel or power costs — although it needs a back-up system, like any other sunpowered heater, to come on whenever the weather's bad. Thermosyphoning solar hot water heaters are generally not as effective as forced circulators in providing a large and dependable hot water supply, but here are some of the very good reasons why people put them in:

• Fluid circulation through the system occurs naturally, without help from a pump, and is completely self-regulating. As long as the collectors can pick up usable solar heat, the heater will keep operating.

• At night, or whenever there's no sun, the circulation stops automatically. This means that no electro-mechanical controls are needed to shut the pump off when the panels can't collect heat any longer.

• Because a pump and thermostatic controls are not necessary, a free-circulating solar water heater is at least $200 less expensive to build than a forced circulator type. (A simple, direct thermosyphon system — including 50 square feet of collector, a 40- to 80-gallon storage tank and plumbing — should cost no more than $500 if you do the installation work yourself.) Operation costs are all but eliminated, too, because no electricity is used. And only rarely is maintenance necessary.

• In a natural circulation system the storage tank must be located *above* the top of the collector panels. For a lot of people this means the tank doesn't have to be placed, as some have said, in the "cold, damp basement," but in the "warm, dry house where it should be anyway." This reduces heat loss from the tank itself — and from the plumbing — and can save extra dollars in insulation costs.

Assuming that there *is* room for a storage tank inside the house, the only considerations in many cases are, "Where do we put the collector panels?" and "Do we have a floor or rafter structure strong enough to support a heavy water tank?"

How Do Thermosyphon Systems Work?

A syphon, as any elementary science student and efficient gasoline thief knows, is a tube that transports liquid up and out of a container at one level to a second container at a lower level — all with the help of atmospheric pressure. As kids we made syphons just by sucking air out of a piece of hose or rubber tubing to create a vacuum and get the liquid flow started (Figure 77).

In the most simple thermosyphoning solar water heater, the main body of water (the storage tank) is always located at the highest point in the system. Downward pressure from the tank displaces water in the cold water tube (which runs out near the base of the tank) and eventually forces it into the channels of the absorber plate, located in the collectors below.

At the same time, the collectors are gathering radiation, changing it over to BTU's of heat that are, in turn, conducted to the water (or fluid)

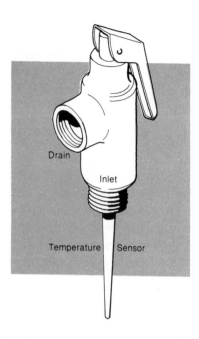

Figure 77. This type of pressure-temperature valve which should be placed at the top of the storage tank automatically opens when it senses too much heat or pressure in the tank. These are normally preset to release at 150 psi.

passing through the absorber plate. So cold water goes into the bottom of the collector and hot water comes out the top (Figure 78).

Heated water, remember, becomes lighter or less dense than cold water, so it rises into the tube that runs back to a point about 2/3 of the way up the side of the storage tank. The phenomenon that causes heated liquid to run upward in a tube — in this case from the collector back to the tank — is called *natural convection*, as you may recall. Heat then, plus a syphoning effect, are the two keys to a free-circulating system (Figure 79).

Meanwhile back at the tank, something called *stratification* is taking place. Water keeps circulating slowly in the tank too, as we already know, but gradually colder water settles to the bottom of the tank while the hotter water remains near the top. So there are several different temperature levels, as there must be in any hot water tank. This why we can draw hot water from the top, and why we can always be sure that cold water will run out of the lower line.

But this stratification is also precisely why the tank *must* be above the collectors. If it were not,

the system would run backwards at night when there was no heat from the sun. If there's nothing to keep the circulation going in the right direction, hot water will rise out of the tank and into the collectors, creating an awkward circumstance known as "reverse flow." Reverse flow not only takes hot water from the tank, it feeds cold water *in*, because the water will actually be cooled as it passes through uninsulated collectors at night. This is why the bottom of the tank should be at least a foot above the top of the collector. Two feet would be even better.

It's been suggested from time to time that a thermosyphon system with a tank *below* the collector could work if a *check valve* (which only permits water to run in one direction) were installed in one of the lines. This is a great idea in theory, but in a typical free-circulating system there's rarely enough pressure in the lines to allow the check valve to function properly.

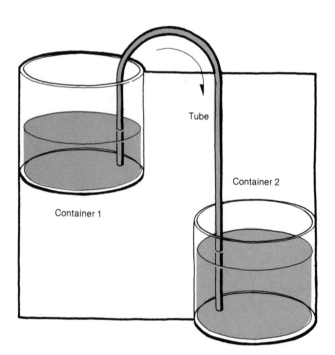

Figure 78. This simple demonstration, which you probably saw in the 5th grade, explains how the downward leg of a thermosyphoning solar water heater works. If the tube is "evacuated" (air is taken out), and container 2 is below the bottom of container 1, normal atmospheric pressure will force water into the tube. As long as the tube stays filled with water, the upper container will drain. This is half of the reason why water (or antifreeze solution) will circulate naturally in a thermosyphon.

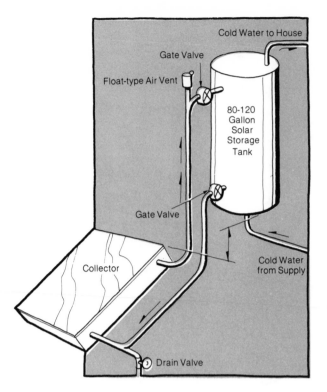

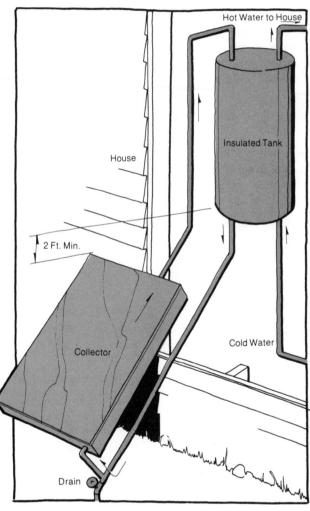

Figure 79. In any thermosyphoning water heating system the base of the storage tank must be at least a foot above the top of the collector. This is to take advantage of the syphoning effect, and to prevent reverse flow when there's no heat coming into the solar collector panel. Notice how the tank is plumbed. Hot water can be drawn from the top of the tank, while cold water comes out the bottom. The gate valves are optional, though a good idea. But the boiler drain cock at the very bottom of the system is critical. Its purpose is to empty the lines during freezing weather.

In a house with a steep enough roof, a tank can be located above the roof-mounted panels — in the attic beneath the ridge pole. (There's advantage in this if the attic is heated and well insulated, because there's less potential heat loss from the tank.) Many people in southern states have disguised their tanks in false chimneys. Another possibility: If your house is higher than your garage, consider putting collectors on the garage roof and the tank somewhere higher in the house. If none of these locations is possible or practical, there's always the ground (Figure 80).

What are the Disadvantages of a Free-Circulating System?

There are some serious drawbacks to thermosyphoning solar water heaters, too:

• For one thing, the hot water-producing capacity is considerably less than in a forced-circulation system. Where water is *pumped*

Figure 80. Where it's not possible to put collectors on a roof and still have the storage tank higher, collectors can be mounted on the ground outside the house. The panels should be located as close to the exterior wall as possible to keep the plumbing lines short. If the tank contains a heat-exchanger coil so the heater can run during cold weather, the plumbing will have to be carefully insulated. Some people protect outside plumbing lines with electrical heat tapes.

through solar collectors, it's reasonable to expect as much as a gallon per square foot of collector surface per hour in ideal conditions. In a natural circulation system, if you're getting more than a gallon of hot water per square foot of collector per *day,* your system is working extraordinarily well. In other words, a thermosyphon system not only starts up slowly, it provides about 1/5 the volume of hot water, even though water that has been circulated naturally will be hotter than water that's been pumped through the collectors at a faster rate.

● A direct thermosyphon system, where no antifreeze is used, presents a problem other than the obvious one of freezing. Because there are no corrosion inhibitors circulating through the heater, impurities in the water can eventually cause mineral deposits to build up in the absorber tubes and plumbing.

● Placing the storage tank above the collectors sometimes means structural problems. It's risky to have your hot water supply above your living space. Even if you reinforce a floor or attic well enough to be sure the whole works can't come crashing down on you in the middle of the night, a leaky tank in the attic can be a real pain in the neck. Repairing or replacing sheetrock and insulation, and having to repaint walls and ceilings that have been ruined by water can cost more than any savings you might make by sticking with a thermosyphon system (Figure 81).

● A natural circulation system must of course be set up so there are no level spots in the lines where airlocks can form. All the tubing must slope either downward to the drain cock or upward to the storage tank. There must also be an absolute minimum of friction within the plumbing itself. In most cases this means using larger diameter and more expensive tubing — perhaps 1 inch or even 1-1/2 inch tubing instead of the normal 3/4-inch soft copper. It also means allowing more space for the plumbing — to eliminate any bends and 90-degree "ells" — and thereby minimizing resistance in the lines.

● Head loss — resistance in the pipes — is enemy number one to any free-circulation solar

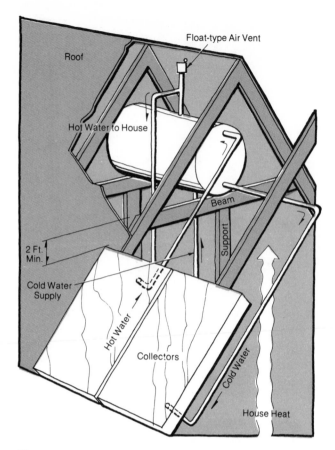

Figure 81. Some storage tanks for thermosyphoning systems can be installed in attics or even in false chimneys. This is a good way to preserve heat in the water — if the tank *and* the attic are well insulated. Don't forget that a 120 gallon water tank can weigh as much as half a ton, so be sure to beef up your roof and floor joist structure.

water heater. The place where there's likely to be the most friction is in the collector itself. Here the channels, either on or in the absorber plate, should be at least 3/4 inch in diameter. And there will be less resistance if the collector tubing is in a series rather than in a parallel configuration. What this means is that you almost have to build your own absorber. Most prefabricated absorber plates have parallel channels that are too small to allow easy water or fluid passage.

● And then there's freezing. The best rule of thumb for deciding whether you can safely install a direct thermosyphon system goes something like this: If citrus fruit *cannot* be grown successfully where you live, forget it! Even if you can

76

grow oranges in your back yard, you still may be taking a chance. This means that in most parts of the United States, thermosyphoning for solar hot water is impractical in a year-round home *unless* the system is set up with a heat exchanger.

How Do Heat Exchangers Work and How Do You Make One?

A heat exchanger is a reasonably simple and efficient device that allows you to do a couple of things: (1) You can keep an antifreeze solution in a closed collector loop that is totally isolated from the domestic hot water supply in the storage tank, as we already know. (2) And *because* a heat exchanger permits two separate loops, your tank can hold whatever normal pressure exists in the home's regular water system (maybe 60 psi) while the collector loop can have considerably less pressure (say, 15 psi).

Remember that heat always wants to travel from a hotter area to a colder place. While this tendency can cause troublesome heat loss from the storage tank, this same principle makes a heat *exchanger* possible. Heat can be passed from a fluid flowing on one side of a barrier to another liquid on the opposite side without the two ever touching each other (Figure 82). The idea is to give the separating wall between the two liquids as much surface area as possible, so that a maximum amount of heat can be transferred.

The most common type of heat exchanger for a solar water heater — and probably the best — is a copper coil immersed right in the hot water storage tank. The longer the coil, the more surface area it has, and the greater the opportunity for heat to pass out of the copper tubing into the water in the surrounding tank (see Figure 68).

In any solar water heater, by the way, the surface area of the heat-exchanger coil should be at least equal to 1/4 the total surface area of the tubing in the absorber plates of the collector panels — *if* the system is to be truly efficient. The coil in a typical 120-gallon-tank heat exchanger is more than adequate to support *several* 24 square-foot collectors.

You can also buy heat exchanger units that can be set up separately outside the hot water storage tank (Figure 83). In this case there must be an extra loop to circulate water from the tank, through the exchanger, and back to the tank by natural convection. This loop must be insulated, of course.

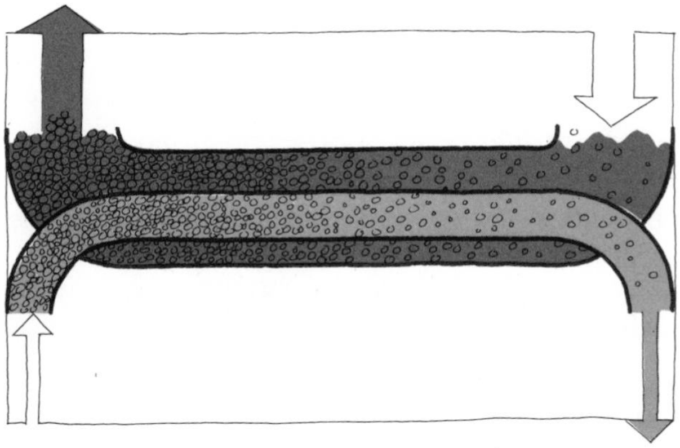

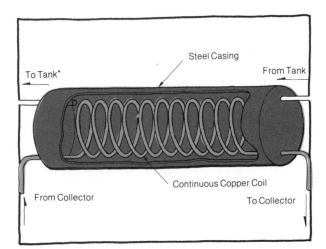

Figure 82. In a counterflow heat exchanger, two liquids move past each other in opposite directions. Heat is transferred from one to the other through the metal barrier. If a coil of tubing is submerged in a water tank, heat is conducted out of the coil into the surrounding water, even though the water in the tank is not moving.

Figure 83. It's possible to buy heat exchanger units that can be installed outside an existing hot water tank. This requires a separate loop to bring water from the tank to the exchanger and back to the tank. Naturally this is not as efficient as an in-tank coil. Any separate unit like this should be heavily insulated. And so should the plumbing.

If a tank and ready-made heat exchanger combination seems like too big an investment, there *are* ways to build homemade heat exchangers that can be used either with thermosyphon solar water heaters or forced-circulation systems.

One of the simplest of these "homemade" heat exchangers can be made from salvaged junk. The design is by Steve Baer and his Zomeworks cohorts in New Mexico. It's no more than an old hot water tank that sits in a recycled 55-gallon oil drum filled with a non-toxic antifreeze and water mixture. Heated fluid enters the 55-gallon "jacket" surrounding the tank and heat is conducted to the water inside. In this plan (Figure 84), the hot-fluid intake for the jacket should be at least 9 inches below the top of the oil drum to insure good heat circulation within the jacket.

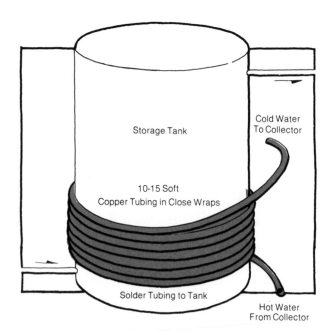

Figure 85. This is another homemade heat exchanger. If you wrap the copper tubing carefully and use lots of solder to sweat the copper to the tank, you will get a fairly good thermal bond. "Thermon" cement should work just as well as solder. Try not to let the copper tubing kink, and be sure to cover both the heat exchanger and the tank with insulation. What you see here is an example of a double-wall heat exchanger — because heat must pass through two barriers.

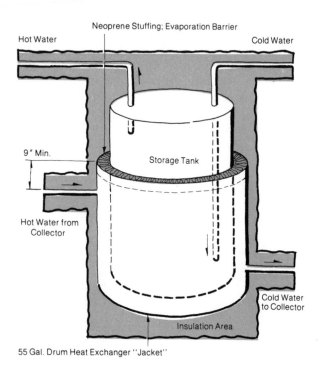

55 Gal. Drum Heat Exchanger "Jacket"

Figure 84. Here is the Zomeworks heat exchanger. It's easy to make, efficient, and cheap. But it should be watched closely. If the wall of the tank ever corrodes through, the domestic hot water supply will be contaminated with antifreeze. Remove the insulation every once in a while to check on things.

Another alternative for an easy-to-make heat exchanger is to wrap about 50 feet of 3/8-inch or 1/2-inch soft copper tubing around the lower 2/3 of the storage tank. Keep the wraps close together and free of kinks. Then use plenty of solder to make a good thermal bond between the tank and the tubing. Once the solder is in place, insulate the whole works. A heat exchanger like this can never be as good as an in-tank coil because there are two barriers the heat must pass through — the tubing itself and the wall of the tank (Figure 85).

Whether you choose a free-circulating solar water heater or a forced-circulation system is, of course, your decision. Even though it has disadvantages, the attractiveness of a thermosyphoning system is its simplicity and its low operating costs, even *with* a heat exchanger. But don't make a choice until you've read the next chapter.

CHAPTER 9

Forced-Circulation Systems

Now let's talk about forced-circulation systems, specifically, in detail. Actually, we've been discussing them all along, since much of what's been said so far leads up to the design in Figure 86. But remember that Figure 86 represents only a *suggested* system. It may be more elaborate (and costly) than you want, and it may not suit your house — either old or new — *exactly*.

It's safe to say that you probably will *have* to make some minor modifications. For instance, you may decide, after reading the last chapter, that you don't need a circulation pump at all, in which case your system will not be "forced." Even if you decide that a thermosyphon system *will* work for you, you may choose to include some of the controls and other gadgets mentioned here.

In many ways Figure 86 represents all of the best elements of sound solar water heating technology at the present time. At first glance it looks complicated and confusing, but it's not — once you break it down and understand what's going on.

Each of the component parts — with the exception of the absorber plate and the special solar storage tank — is a standard, time-tested plumbing element that should be sitting on a shelf in your local plumbing supply outlet. There's no more need to lose sleep over the possibility of your system's failing than there's need to worry about a breakdown in your regular household plumbing. That's because basically a forced-circulation solar hot water heater *is* regular household plumbing.

To keep things simple, we'll divide the heater into its three major parts: (1) the collector loop, (2) the hot water loop, and (3) the thermostatic controls. Then, we'll trace the flow of antifreeze fluid from the storage tank (beginning at the very base of the collector loop), to the collectors and back through the heat-exchanger coil. Next, we'll follow the path of the water, beginning at the cold water supply, as it goes through the solar storage tank and existing hot water tank, to the house itself. Finally, we'll explain how the control system operates.

What's in the Collector Loop Besides a Pump?

The whole collector loop (start #1, Figure 86) should consist of 3/4-inch Type L soft copper or Type M rigid copper. You'll also need standard 3/4-inch sweat-type fittings, valves, threaded nipples, and other fixtures throughout. These should be joined together with high-grade 95-5 solder which withstands heat better than regular 50-50 solder. Any threads should be coated with Teflon pipe dope normally used in refrigerator tubing. Standard plumbing compound will dissolve in propylene glycol, and threaded joints will quickly leak.

The lowest point in your system no doubt will be where the cold water tap leaves the solar storage tank. Here there should be a short section of tubing and, of course, a boiler drain which can be used to empty the collector loop quickly. This is also where you will most likely fill the system with fluid for the first time, so use a drain cock with hose-connecting threads.

Somewhere in the lower part of the loop — either before or after the pump — there should be a poppet-type pressure relief valve (like the Watts model #174A), set to "pop" at 30 psi (Figure 87). If this *should* blow for some reason, it will dump antifreeze fluid on the cellar floor, so it might not be a bad idea to keep a bucket under that valve so you can save any fluid that's released.

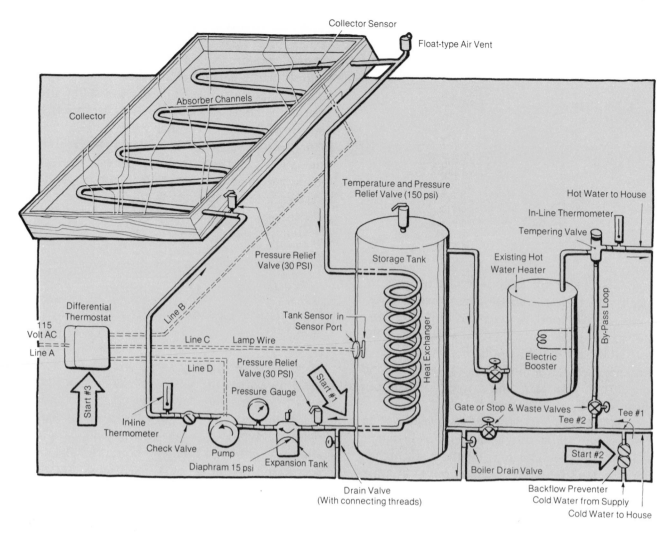

Figure 86. Don't be frightened by this diagram! It's not as complicated as it looks. Begin at any one of the three starting points and trace the system through slowly. The collector loop (#1), the hot water loop (#2) and the control system (#3) are all explained in the text.

This pressure relief valve shouldn't cost more than $8. But keep in mind that the best ones are not adjustable, meaning that you'll have to buy one that's pre-set to release when there's 30 pounds per square inch of pressure or more in the lines. This relief valve, by the way, might be only the first of several in your collector loop. Plan others for any *high* points in the system up above.

Also somewhere low in the loop — where you can get at it — there should be an expansion tank and an *air purger* (like Amtrol's #443) with an automatic air vent (like Amtrol's #701). Excellent expansion tanks are sold by Amtrol (model #15), Heliotrope General and others (see appendix). Don't spend more on one than about $22.

Inside the expansion tank, where you can't see, there's an airproof rubber diaphragm with its edges fastened all around the sides of the tank. Pressurized air, pre-charged at 15 psi, is below the diaphragm, while fluid from the lines runs through the air purger above the tank. As the fluid heats and expands, the increased pressure from the expansion is taken up at the flexible diaphragm (Figure 88).

This is how the system is able to maintain constant pressure at 15 psi. To check on this you may want to put in an inexpensive pressure gauge ($5) somewhere near the expansion tank. An additional option might be to install in-line thermometers at various points in the loop so you can monitor what's going on with the tem-

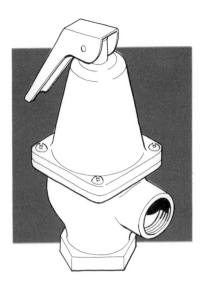

Figure 87. Here is a typical poppet-type pressure relief valve. A safety valve like this will blow when there's too much pressure in the lines, caused by water expansion. The best pressure relief valves are spring-loaded and are not adjustable. You'll want ones that are pre-set at the factory to release at 30 psi.

perature. A *tridicator,* a gauge which measures temperature, pressure *and* altitude, is a more expensive possibility. If you buy one of these you'll be saving a tee fitting or two.

Now the pump: The flow rate through a solar collector loop is normally very low — about 5 gallons per minute — so you don't need a very strong pump. A 1/20- to 1/12-horse-power (HP) circulation pump with a cast iron housing should be more than powerful enough if everything else is set up properly (Figure 89).

Whether you choose a 1/20 HP or a 1/12 HP model will depend on two factors: (1) how high you must pump fluid to the collectors, and (2) the amount of head loss in the loop. (Head loss is measured exactly with a "manometer," which you will have to rent or borrow from a plumber.) This rule may help your decision: *Generally, if you're going to pump fluid to an elevation of two stories or less — and have fewer than 6 or 7 collectors — the smaller pump, costing about $61, is adequate.* Pumping to many collectors at 3 stories or higher, you'll need the 1/12-HP pump, priced at about $69.

In either case, be sure you buy a pump that won't be corroded by the antifreeze fluid. If some

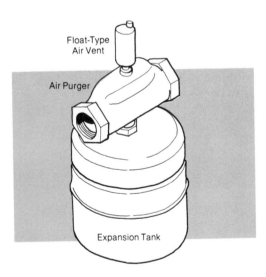

Figure 88. When a fluid containing water is heated, it accumulates dissolved air, which can collect and form bubbles that block the liquid's flow inside the tubing. Above is a diaphragm-type expansion tank in combination with an air purger. The tank maintains constant pressure in the lines by taking up the fluid expansion. At the same time, more dense fluid flows through a lower passage in the air purger. Lighter fluid, containing air, moves through an upper channel. Here air is freed from the liquid, and released through an air vent. See the illustration of an air vent in Figure 90.

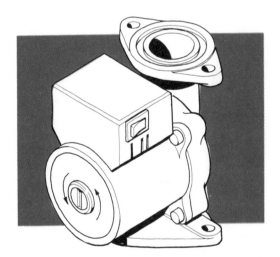

Figure 89. A circulating pump like this can be installed either horizontally or vertically, but it must never be put in upside down, or it will run backwards. Good pumps like this one have a variable speed adjustment, and two different motor speeds. Put isolation valves on either side so the pump can be serviced without having to drain the whole collector loop. These valves come with the pump.

of the pumping parts are aluminum, for example, the pump will die an early death. As far as we know, the Grundfos Corporation of Clovis, California makes the best and most reasonably priced circulation pumps for our purposes here — their models UPS/UMS 20 and UP 26 (see appendix). They run on ordinary house current (115V AC). They also have a variable adjustment for flow control — which is controlled by a handy external adjustment — and 2 different motor speeds. (It's also a good idea to *use* the isolation valves that are provided. Put them on either side of the pump, so they can be closed off if you ever have to do some pump maintenance.)

To keep warm fluid from rising out of the tank at night when the collectors are cold (this creates a reverse flow, remember), there should be a check valve somewhere in the lines. The little "flapper" in a check valve — which allows water to flow in just one direction — only works right when the valve is installed in a horizontal or near-level line. The most convenient place to put a check might be on the same level with the pump. If you don't have a handy level stretch of line, put in a *flow-control valve,* which can be installed vertically.

So fluid comes out of the tank, past the drain valve, through the pressure relief valve, the expansion tank and the gauges, and is pumped through the check valve into the line of tubing leading to the collector. There it flows upward through the channels in the absorber plate and gets heated by solar radiation.

The line coming *out* of the collectors should run level or even slightly uphill to a float-type vapor relief valve like the Amtrol #701 ($8 to $12), which releases trapped air (Figure 90). Any air bubbles in the lines that have not been "purged" at the expansion tank should rise to this air vent, displace the fluid next to the float, and escape out the top. This type of valve not only "bleeds" air out of the system, it also provides vacuum relief, letting air into the lines when you drain the system. But çaution: Float-type air vents are delicate. If the highest point in your system is outdoors and unprotected, the valve can be snapped off by snow and ice build-up on the roof. So protect it as best your can.

From there it should be all downhill — literally. (Don't leave any hollows or dips in the return line

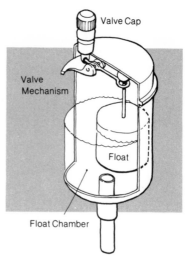

Figure 90. A float-type air vent like this should be installed at all high points in the system. It works like this: when an air bubble from the line comes into the chamber, the water level is lowered. As a result, the float drops and opens the air valve. As the air escapes through the valve opening, the water level rises again, and the valve closes. The valve cap should be opened about ½ turn to permit air to escape.

where air might be trapped.) The hot antifreeze and water mixture passes back down the line to the heat exchanger coil in the storage tank, runs through the coil (giving up most of its heat), and comes out cool again at the base of the tank.

So we're right back where we started — ready to have a look at the separate heater loop.

What's in the Hot Water Loop Besides a Heater?

Cool, street-temperature or ground-temperature water (start #2, Figure 86) comes from outside the house through the cold-water supply line. As soon as it enters your system there should be a device known as a *backflow preventer.* This is nothing more than a double check valve with an atmospheric vent between the two checks. It doesn't allow water to run back out of your system (Figure 91).

If for some reason your tank or heat exchanger were to fail and there were a contaminating leak

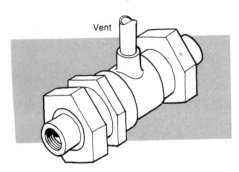

Vent

Figure 91. Here is what a backflow preventer looks like. It has two flapper discs — which don't allow water to pass back out of your system — and a vent. These devices are required by law in many places.

between the two loops of the solar water heater, no poisonous antifreeze could escape into your well or the public water supply — although your house supply *would* be contaminated. Backflow preventers (like the Watts #9D) are required in many municipal systems, anyway. Even if it's not mandatory where you live, it's a good safety precaution to have one.

Once it has passed through the backflow preventer the water should reach a tee. Here some will be diverted to supply the house with cold water, and some will move on toward a second tee. At this second intersection (Tee #2 in Figure 86), part of the remaining water will be directed *around* the solar storage tank and the existing heating element to later be mixed with hot water coming out of the tanks. This by-pass meets the main hot water line at a tempering valve, which *we'll* approach in our discussion from the other direction.

Water that's to be heated passes through an open gate valve and past the tank's boiler drain into the storage tank. (Cheaper stop-and-waste valves can be used instead of gate valves in the heater loop. These have the advantage of a small thumb-screw drain, but water has to follow an indirect path as it passes through a stop-and-waste valve, meaning there's lots of friction there. This much head loss is all right if there's plenty of normal pressure in the lines — 60 psi, for example. But in the collector loop, which has only 15 pounds of pressure, this much resistance in a valve will only cause the pump to work harder than it should have to. Fluid will flow straight

through a gate valve, on the other hand, so use *only* gate valves in the collector loop.)

Once it's in the tank, the water circulates gently, picks up heat from the heat exchanger, gradually rises to the top of the tank and out the tube leading to the existing hot water heater where its temperature will be boosted if necessary. This connection between the solar storage tank and the domestic hot water heater should also have a gate or stop-and-waste valve in case you need to shut down part of the system.

If it's a sunny day, the water should have picked up enough heat so there's no need for the oil-fired, gas or electric heating element to switch on. If not, the water will be heated some more to satisfy the heater's thermostat, be stored, and be drawn off as needed. When someone turns on hot water in a bathroom, very hot solar-heated water will move out of the heater tank toward the tempering valve, where it joins the cold water from the by-pass line.

A *tempering valve* (like the Watts #70A) is nothing more than a mixer (Figure 92). It has a small spring-loaded adjusting knob on its top which permits you to regulate the temperature of the water going into the house hot water lines. By mixing hot and cold together, this little device — costing about $12 — will adjust water temperature between 120 degrees and 160 degrees Fahrenheit, depending on where you set it. (Right

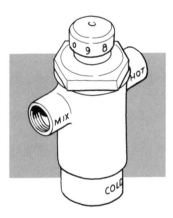

Figure 92. A tempering valve like this one mixes hot and cold water before it goes into the house lines. The knob on the top lets you dial a desired temperature. To raise or lower the water temperature, anywhere between 120 degrees and 160 degrees, just adjust the knob.

after the tempering valve is a good place for one of those in-line thermometers, so you can check on the water temperature at the valve. Some tempering valves have no precise temperature settings, just a little dial numbered from 1 to 10.)

A tempering valve is a nice option — although it's not vital. It prevents too-hot water from eventually coming out of a faucet to burn kids' hands. It also keeps more heat in storage because hot water is being mixed close to the tank rather than at the faucet.

After it's past the tempering valve, the solar-heated water passes through normal channels until it reaches the shower head and then runs down your back, where it's yours to relish.

How is All of This Controlled?

You are the master controller because you set the temperatures and determine the pump speed (start #3, Figure 86). But when you're not around to oversee things — which is most of the time — a small electrical box with a differential thermostat takes charge. Out of this box run 4 electrical lines. *Line A* plugs into a regular wall socket. *Line B* leads to a sensor in the collector's absorber plate. *Line C* goes to a sensor against the side of the solar hot water tank, and *Line D* leads to the pump.

Electricity runs through the differential thermostat to the wire leading to the *thermistor sensor* on the absorber plate. This sensor should be located on the absorber as close to the outlet of the collector as possible, because its job is to know how hot the fluid is as it comes out of the panel.

Meanwhile, there's a second sensor at the end of a second wire circuit. It has been installed in the sensor port, located part way up the side of the storage tank. This second sensor's job is to know the temperature of the water in storage. (Both end sensors should be mounted with silicone fiberglass tape, or better still, liquid aluminum epoxy, so there's good thermal contact.)

Each of these sensors — the one in the collector and the one in the tank — is about 4 inches

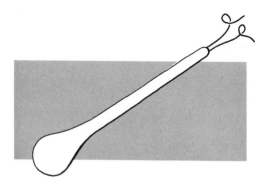

Figure 93. *Thermistor sensors* look like this. They're about 4 inches long and have a tiny electrical resistor in one end. This inexpensive part is small, but it's critical to your solar water heater's control system.

long and has a tiny electrical resistor in it (Figure 93). When there's more heat in the resistor, there's more resistance to electricity passing through it. Likewise, when there's less heat, *more* electricity flows back through the return wire to the thermostat. This is how the thermostat "knows" the water temperature in both places.

When the sensor in the collector tells the thermostat that fluid there is, say, 10 degrees hotter than the fluid below in storage, the thermostat sends an impulse to the pump. As soon as the pump receives this electrical message, it comes on — circulating fluid through the collector loop. When the sensor mounted against the side of the storage tank says the temperature there is getting close to the temperature in the collector, the pump is instructed to turn off (Graph 4).

In other words, the antifreeze-and-water fluid circulates whenever the temperature in the collectors is hotter than the water storage temperature by whatever preset amount you choose. (A 10-degree temperature differential to start the pump, and a 3-degree differential to stop it, are typical settings.) That's how the control system works. It's pretty foolproof.

A slightly more expensive variation on a simple differential thermostat has a feature that automatically turns on the pump whenever the collector temperature gets lower than 36 degrees Fahrenheit. This causes warm fluid to circulate through the panels, to prevent snow build-up, until the collector temperature reaches 37.5 degrees. Then the pump shuts off again, because the collector is warm enough to melt any snow

Graph 4. The heavy lines in this graph chart two different temperatures in a solar hot water heater during a 24-hour period. One line represents the collector temperature, the other the temperature of the water in the storage tank. Here's how the day went:

1. At about 8:30 in the morning the absorber plate in the collector got enough warmer than the tank water to allow the pump to turn on.

2. Notice how the collector temperature took a quick dip as soon as circulating fluid started to cool it off. You can see that it recovered quickly.

3. At about 5:00 that afternoon, the collector temperature dropped to within a few degrees of the water in the tank, as the sun got lower in the sky. At this point the differential thermostat instructed the pump to shut off.

4. Because there was no longer any fluid circulating

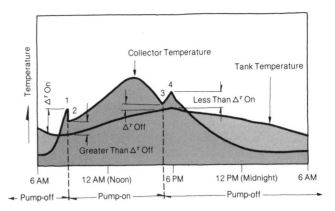

to cool the panel, its temperature rose slightly. But there was not enough temperature differential (**DT**) to start the pump a second time. It stayed off for the rest of the night, as the collector cooled well below the temperature of the water in the tank.

that lands on it. (A differential thermostat like this costs about $40, as compared to $35 for a regular one.)

Another more sophisticated thermostat has an "automatic off" which doesn't allow the pump to turn on if the temperature in the collectors falls below 80 degrees F. (This will prevent the system from fruitlessly trying to collect heat on a warm night after you've used up all your hot water in storage.) The same device can also be set to refuse electricity to the pump whenever the storage temperature gets above, say, 160 degrees.

Any of these differential thermostats should, of course, have a manual override. The cheapest and best ones come from companies like Rho Sigma and Heliotrope General, both of California (see appendix).

The slower the fluid goes through the collector, you'll remember, the hotter it gets. But you should also recall an earlier lesson, that a collector is more *efficient* when more fluid is running through it at a faster rate. You can make your whole system more productive simply by adjusting the speed of the pump. On a warm sunny day, you may set it to run faster; when there's haze, it should run slower. Your in-line thermometers will help indicate what to do — *if* you

decide that varying the flow rate is necessary at all.

There *are* some special variable-speed pump and differential thermostat arrangements that coordinate with a "rheostat" unit. A rheostat works like a dimmer-switch for lights. When the collector sensor indicates lots of heat in the panels, the rheostat feeds more "juice" to the pump and circulation speeds up. When there's less heat in the collectors, the pump gets less electricity and slows down again. These are rather sophisticated and hard to hook up. Besides, there are very few thoroughly tested systems of this type available at this time. One we do know about is made by Rho Sigma. Contact them for more details. If that seems like too much trouble, stick with manual speed control of your circulation pump — or none at all.

That's all there is to it. It wasn't as complicated as you thought. We still bet there's no reason you can't put this whole thing together yourself, once you have the right parts. To be on the safe side, though, sketch your intended system on paper and show it to an experienced plumber. He'll be able to point out any obvious goofs and weaknesses immediately.

CHAPTER 10

Solar-Heated Swimming Pools

We've saved our discussion of solar-heated swimming pools for last — not because it's least (*or* most) important, but because it's different. On one hand, swimming pools are still considered luxury items, even in the minds of most affluent Americans. Domestic hot water, on the other hand, is something we take for granted in most homes in the United States. To put it another way, we see hot showers as one of life's necessities, but heated swimming pools as expensive frills.

As we get more and more strapped by our dwindling supply of traditional energy sources, the public will disapprove more and more of burning precious fuels to heat pool water. Already legislatures in California, New York and Illinois lean quite specifically on pools that have gas-powered heaters. And laws against nonessential consumption of fossil fuels will only get tougher as time goes on.

If you already have a heated pool, or want one in the worst way, heating your water by the sun is an easy — and acceptable — way out. You see, you have two things working for you before you even start. In the first place, the pool itself is a natural solar collector. Second, it's a ready-made solar storage tank.

When you come right down to it, what most people are trying to do with any pool heater is just raise the temperature of the water to about 80 degrees Fahrenheit, and then *keep* it there. The idea, usually, is to extend the swimming season for several weeks during the spring and fall — and to be comfortable. In light of what we've already learned about solar potential, it doesn't seem that hard to collect enough free BTU's to heat a pool that much. And it isn't.

On a bright sunny day in summer, a typical pool will gain 2 to 3 degrees of temperature on its own — without help from *any* artificial heating system, solar or otherwise. If the surrounding air temperature stays reasonably high at night, so there's not a lot of heat loss, the pool will continue to collect and store BTU's of solar energy during the day. This process will continue until the water is about 10 degrees warmer than the ambient air. At this point, it can't go any higher, because it reaches a state of equilibrium: it loses as much heat at night as it gains during the day. In other words, unless it has some kind of heating assistance — or some sort of insulating cover — the water in a swimming pool can almost never get warmer than 10 degrees above the surrounding air temperature.

How Much Do Swimming Pool Covers Help?

An opaque or transparent swimming pool cover — which will transmit solar radiation to the water — can give your pool *another* 10 degrees of natural storage capacity. This means that if a free-floating cover with pockets of trapped air is stretched over the pool at night (and whenever no one is swimming), we can expect the water temperature to stay at up to *20* degrees above the surrounding air (Figure 94).

Insulating "pool blankets," as they're called, pass about 85 percent of the light that hits them through to the water. Some are even mounted on convenient reels, which let them be stretched out and rolled up again without a lot of hassle. Ideally, of course, a pool cover should be custom-fitted around the edges, so there's an absolute minimum of heat loss resulting from evaporation, re-radiation and air convection. (A wall, hedge, fence, or wind screen around the perimeter of the pool, by the way, is another potential heat — and money — saver. This will help to keep passing wind from robbing heat from the water.)

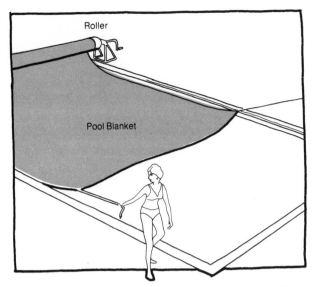

Figure 94. A swimming pool is a natural solar collector and storage area. But we tend to take an easy-come-easy-go attitude toward free solar heat, letting a lot of it escape back to the atmosphere. Air-filled pool covers, which let light through to the water, can hold heat in a pool and reduce the need for additional heating. The cover, or "blanket," should stay on the water whenever no one is swimming. That's why it's good to be able to spread and remove it easily.

One of the finest insulating pool covers is manufactured by L.M. Dearing Associates, 12324 Ventura Blvd., Studio City, California 91604. In the eastern United States the Dearing Solar Pool Blanket is distributed by the Sealed Air Corporation of Fairlawn, New Jersey 07410. There are also other fine pool covers. Ask a local swimming pool dealer what he recommends.

There are also some bizarre and interesting alternatives to covers, which have to be taken up and laid back down on a pool. One is plastic air-filled "lily pads" that float on top of the water and hold the heat in. Then there are the many-colored spheres that look like big ping-pong balls. In either case, you order enough of these to cover the whole surface of your pool.

Then you can swim, and dive into the water, almost as if they weren't there. These balls seem a little freaky at first, but they're fun, provide pretty good insulation, and are no trouble at all once they're in place. (What's *most* astonishing about the balls is that once you get out of the pool, and let the water settle down, they will naturally line themselves up in perfectly straight rows and stay

that way until the water is disturbed again — regardless of the shape of the pool.)

One thing more: Believe it or not, in many cases there is already a beautiful solar collector right next to the swimming pool. We're talking about the concrete deck. (Your feet know how hot this can get when the sun is strong.) If it were painted a dark color, it could collect an amazing amount of heat from the sun.

Some homeowners have designed portions of their concrete slabs to drain back into the pool itself. In this case, water is pumped out of the pool and through the filter — as it normally is — but then gets sprayed gently and evenly onto the darkened deck surface. From there it flows slowly — and in an even film — back to the pool, picking up heat all the way.

A small architectural engineering firm in Vermont has developed a pool heating system along these same lines. They see the whole pool deck as a flat-plate collector — literally. The steel-reinforced concrete slab sits on top of gravel and is insulated against the ground by sheets of styrofoam. The top of the slab is painted black, deep green, or some other dark color that will absorb radiation (Figure 95).

When the slab is warm enough, water is pumped out of the pool and diverted through a series of 1-inch plastic pipes imbedded in the concrete slab about 10 inches apart and 4 inches below the surface. Needless to say, the concrete isn't as efficient as a copper absorber plate, but it's more than enough to keep swimming pool water at 82 degrees throughout the season.

Can Pool Water be Heated with Solar Collector Panels?

For some reason, cheaper solar collectors designed specifically for swimming pools are called "heating panels" rather than "collectors" — even though they work on the same principle. Regular collectors — the type we've been discussing throughout this book — *can* be used. But it's a little like taking a Ferrari to pick up a load of 2 x 4's at the lumber yard.

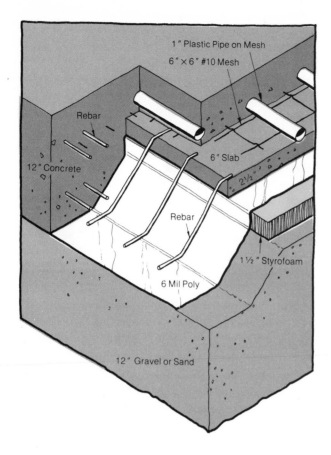

1" Plastic Pipe on Mesh

6" × 6" #10 Mesh

Rebar

12" Concrete

6" Slab

2½"

Rebar

1½" Styrofoam

6 Mil Poly

12" Gravel or Sand

Figure 95. Using a little imagination, designers at Parallax in Hinesburg, Vermont, have built a concrete pool deck that doubles as a flat plate collector. There are several features you should notice:

1. The concrete slab sits on at least a foot of well-compacted gravel or sand.

2. The slab is "floating," which means that it's not firmly anchored to the ground anywhere. This is why the edges of the concrete "turn down," or get thicker. The turn-downs keep the deck from shifting.

3. The insulation on the bottom of the slab is 1-1/2 inch styrofoam.

4. Between the insulation and the slab, there is a vapor barrier. This is no more than a sheet of polyethylene.

5. The slab is held together and reinforced with steel bars running in both directions — as well as #10 steel mesh.

6. The water channels — 1 inch plastic pipes — sit about 4 inches below the surface of the deck, which is painted dark green. Filtered water is pumped through these channels before it returns to the pool.

This type of solar pool heater is about 45% efficient — more than adequate to keep pool water warm.

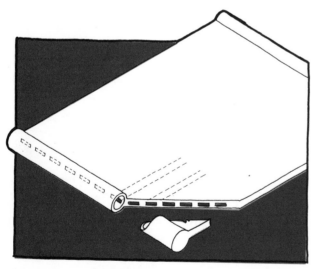

Section Through Solar Panel Showing Fluid Passages

Figure 96. One of the best known and most successful pool heating solar panels looks a little like a black windowshade. The water passages and headers are extruded right into the plastic, so the whole panel is a single piece.

Large-capacity pool panels, unlike high-performance collectors for domestic hot water, operate in the low-temperature range. Obviously there's a big difference between heating a fluid to 160 degrees or higher on a winter day, and simply warming it to 82 degrees on an already-warm day. You don't need anything very fancy to make pool water tepid (Figure 96).

Today the market is loaded with swimming pool panels. Some of them are good, and some not so good. Sometimes they're made of plastic, sometimes of rubber, and sometimes of metal. Most have no glazing because they work when the ambient air temperature is high and because water circulating through them doesn't need to get that hot.

There are some problems with some of them: Ultraviolet rays, for instance, can raise the dickens with certain types of plastics that contain no ultraviolet inhibitors. And ozone will eventually break down certain components in rubber. After some of these cheaper panels have been used for a while, they end up looking like a blackened potato chip — curled, warped, chipped and cracked.

Watch out for aluminum panels, too. Chlorinated water corrodes aluminum very quickly. Again, copper is much safer. Some of the better panels have copper tubing covered by an aluminum absorber plate. They work fine for heating pool water, but, like all other pool panels, they'll never be suitable for domestic hot water applications (Figure 97).

Be careful what you choose. There are too many questionable manufacturers in this new and lucrative field. In some parts of the country the Better Business Bureau is actually warning prospective buyers against fly-by-night installers and distributors of solar pool heaters — all of whom claim their products are made with the "right" materials.

Check out the reputaion of any company you're thinking of dealing with. Find out how long they've been in business, and see what kind of warranty they offer. And don't swallow any extrav-agant claims like, "Our system will give you comfortable year-round swimming in any climate." That's an open invitation to a shady deal.

One of the oldest and most respected names in pool heating systems is Fafco, located at 235 Constitution Drive, Menlo Park, California 94025. Fafco has dealerships in many parts of the country. But Fafco is not the only one; there are plenty of other superior pool panel makers around, too (see appendix).

A complete solar pool heating system — with panels — shouldn't cost more than $1000 to $1500, fully installed. (It could be less, of course, if you do your own plumbing.) So what can you expect from all this? In Florida, where it's sunny and warm most of the year, an extensive solar heating system for a 10,000 gallon pool — costing as much as $2500 — *might* keep the water at 75 degrees even between October and April. Even there, some auxiliary heating might be needed to keep the water very warm. In short, if you live in Atlanta or points north, don't rely on the sun to keep your pool warm all winter.

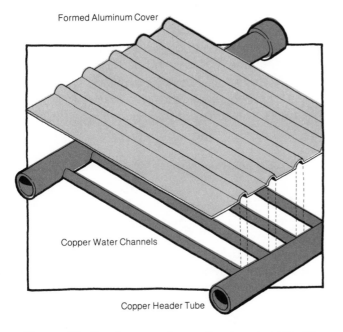

Figure 97. Aluminum makes a fine inexpensive absorber pláte, but aluminum tubing will corrode when it's in contact with chlorinated water. At least one manufacturer puts an aluminum absorber on top of copper tubing, which cannot corrode. This is an excellent combination for a solar pool heater.

Formed Aluminum Cover

Copper Water Channels

Copper Header Tube

How Are Solar Pool Heaters Set Up?

The roof of a pool house or cabana makes an ideal location for solar heating panels — especially if the roof faces south. If there's a flat roof, or if you decide to mount panels on the ground, they will have to be tilted a few degrees to give you reasonable heating efficiency. You may want to build a rack that raises the tops of the panels so they face the sun at a better angle for your latitude — even though finding the *exact* angle is not critical. Any structure that supports the panels should be super-strong and anchored to the roof or ground so the wind can't *budge* it.

There should be no need to insulate either the water lines or the backs of the panels, but any nails, screws, bolts or lags that fasten the panels should be caulked to keep the roof from leaking. (A flat tar and gravel roof may have to be refinished in spots with hot tar.) It's particularly important to have the panels secure where the feed line and return line to the pool attach, so no joints can be broken.

How many panels will you need? That depends mostly on how big the pool is. The normal rule of thumb is to have a total panel surface equivalent to at least 1/2 the surface of the pool (that's if the panels face south or slightly southwest, and assuming that they're more or less properly tilted). If the panels face east or west, or if the panels are lying flat, you may need a panel array equal to 70 to 75 percent of the pool surface.

But experience has shown that this formula provides more heating capacity than you really need in most cases. Remember: Once you get your pool water to a comfortable temperature, the panels won't do much except balance off your nighttime heat losses. Less than 50 percent is usually enough panel area, although with fewer panels it may take as long as 4 or 5 days to heat the pool 10 degrees above the ambient air. But once you get there, this 10-degree gain should be easy to maintain during the summer, spring, and fall, *especially* if you have a pool cover.

Pool heating panels, like all flat plate collectors, work best when there's lots of water running through them. If you keep the temperature differential between the pool and the water in the panels as small as possible — by running the system a lot — it will be most efficient. Water should return to the pool from the panels no more than 6 degrees warmer than the pool itself. If it's a lot hotter than this, you're not operating at maximum efficiency.

Because a pool system must have a large circulation capacity, and because pool panels operate at such high efficiency (70 to 80%), you may want to use 1-1/2- to 2-inch water mains. And be sure your pool pump is strong enough to keep a large supply of water going through the panels to keep them cool. Most pool pumps are more than adequate.

As you plan your system, try to keep the lines between the pump and panels as short as possible. You don't need to worry that much about heat loss, the way you would with a domestic hot water system, but you *are* concerned with keeping head loss at a minimum. The shorter the lines, the less resistance in the pipes. Here are some other hints to keep in the back of your mind:

1. Make sure your unglazed panels are self-draining. This is not so much to prevent freezing, as it's a fail-safe mechanism. If for some reason the pump should fail, water must automatically run out of the panels, eliminating any possibility of pressure build-up and damage to the system. Because there's no glazing, there's no danger of damage to the panel when there's no water in it. Remember too that any self-draining system *must* have an air vent near its top or it will not drain.

2. Keep your pool filter clean. Don't forget that you don't have a closed-circuit collector loop like the one in a domestic hot water heater. There's plenty of opportunity for debris to fall into the pool and work its way into the panels. Once there, it could destroy them. A fouled strainer and clogged filter can bog down the whole system, put unnecessary strain on the pump, and slow the progress of water through the panels.

3. Use plastic plumbing. Polyvinyl chloride (PVC) or chlorinated polyvinyl chloride (CPVC) is much cheaper than comparable copper plumbing. It can be cut in an ordinary carpenter's miter box, and be glued together with regular CPVC cement. Plastic pipe looses some of its strength at temperatures above 180 degrees Fahrenheit, and this is why it's not recommended for regular domestic hot water. But for a solar heater that operates well below 100 degrees most of the

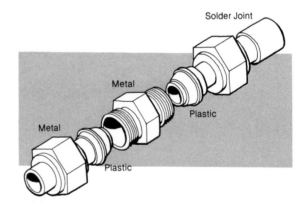

Figure 98. Special adapters will be needed if you use plastic plumbing in your solar pool heating system. Ells and tees will be no problem, but certain valves are only made in metal, and these cannot be connected to plastic directly.

time, plastic will hold up fine. Valves and special fittings may have to be metal. When you put these in, you'll need special (but cheap) metal-to-plastic and plastic-to-metal adapters (Figure 98).

How Does a Solar Pool Heater Work?

In a fairly sophisticated solar pool system, water is pumped out of the pool, through the strainer, and through the filter. After the filter it passes through a check valve. This check valve is more than just an option. Its purpose is to keep water that's draining out of the panels from back-washing the filter (Figure 99).

Once it has passed through the one-way check, the pool water reaches a critical tee. This tee separates the upper panel loop from the ordinary circulation loop. When the pool temperature is sufficiently warm — or when there's no sun to heat the panels — water circulates only through the filter loop. A gate valve in this lower loop, immediately after the tee, determines which way the water will flow (Figure 100).

Whenever this gate valve is *closed,* water is forced upward through the panel loop to collect heat. When it's open, there's normal circulation through the lower loop. This gate valve can either

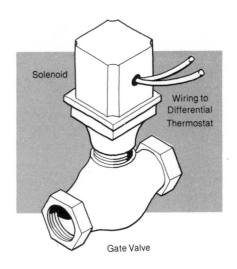

Gate Valve

Figure 99. An electronically controlled valve like this one makes an automatic solar pool heater possible. When there's heat in the panels, this valve is instructed to close. And when the pool water is hot enough, it opens again, circulating water only through the strainer and filter. (See appendix.)

be operated manually or controlled by a differential thermostat such as the one mentioned in the last chapter (Figure 101, next page).

Water moves up through the panels and comes out again — we hope a few degrees warmer than when it went in. Near the top of the

Figure 100. Here is a more detailed look at the critical tee that separates the two circuits in a solar pool heater. When the gate valve is closed, water is forced into the panel circuit. When the gate valve is open, water passes only through the lower loop.

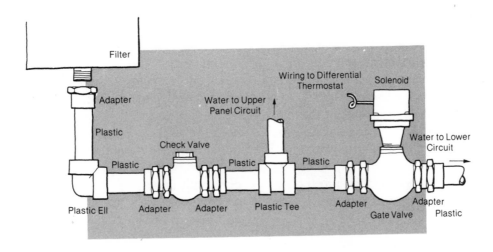

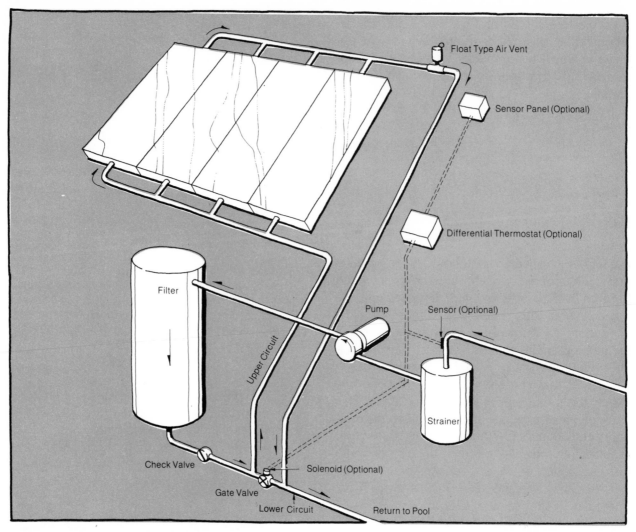

Figure 101. A solar pool heater has two distinct water circuits: (1) the lower filter loop, where water circulates all the time the pump is operating, and (2) the upper panel circuit, where the water picks up heat. The gate valve — which can be operated manually or controlled by a differential thermostat — determines which route the water will take. Note that the thermostat has no control over the pump, but that it *can* open or close the gate valve. The special sensor near the panels is sensitive to both sunlight and heat.

system it will pass through that same small, float-type air vent that appears in a forced circulation domestic solar heater. The panels and panel loop will be draining and filling constantly, and the air vent will allow air to escape from the lines each time the upper loop is filling as well as let air *into* the lines when that loop is emptying. From the panels the heated water simply circulates through the rest of the upper loop, into the last part of the lower filter loop, and back to the pool again.

The all-important gate valve is opened or closed depending on the temperature of the water and on the weather. Many people do it by hand. *If* you want a system that works automatically, you can stick in a remote-control "solenoid" valve that can open and close at a command from a differential thermostat.

A common circulation pump for a pool works constantly unless it's either switched off manually or controlled by a preset timer that shuts it off and turns it on again at certain times of day. So the differential thermostat, which controls the upper loop, is not wired to the pump at all. The pump just keeps running.

But a sensor in the line leading out of the pool to the strainer, measures the temperature of the pool water. If the sensor says the water's 82 degrees (or whatever temperature you want it to be), the solenoid-operated gate valve stays open and no water goes to the panels.

When the pool temperature falls below 82, and a special sensor box near the panels indicates that the temperature and sunshine there is strong enough, the differential thermostat tells the gate valve to close, and water goes into the upper loop to be heated. Whenever the two sensor temperatures get within a few degrees of each other, the valve opens and the panels are by-passed.

That's how it works. Simple. Enjoy. Oh, one last thing: Don't forget about the sun. It's always there, and when it shines, it's powerful and useful, as we've seen. It can also give you a wicked burn. So keep your shirt on. Especially when you're mounting panels.

Afterthoughts

Everyone seems to be thinking, reading, writing and talking about solar energy these days. That's good. News about solar technology — like sex — pops up regularly in almost every newspaper and magazine in the country. It looks as though both sex and the sun are here to stay. And we can be thankful for both.

But just about everything we pick up starts out the same way: "We have too many people . . . . We have this terrible energy crisis . . . . We've been in an 'our-energy-will-last-forever' frame of mind for too long, and so we're running out of fossil fuels . . . . We've got to find ways to use alternative energy sources or we're all going to die (or at very least, drastically change our style of living) . . . . The sun's the answer to all our woes."

That's the bad-news approach — scare tactics followed by unrealistic promises and pie-in-the-sky predictions. Most of us already *know* that we have very little fuel left, and we're frightened enough already. So we don't need more of that. Our energy situation *is* scary, there's no doubt about it, but the idea that all of our energy needs can be served by the sun is mostly science fiction. We'll have to keep exploring other things, too.

The sun *does* seem to offer many kinds of possibilities. It's always there; it appears to be a boundless source of energy, and no one expects that it's going to disappear in the next several million years or so. In other words, it seems to be a perpetually renewable resource.

What's even more encouraging is that we already know a great deal about how to harvest energy from the sun. But we still need to figure out ways to store it effectively for longer periods of time than we can now. That's the hang-up. As we all know, sunlight is not available all the time. So there are some basic and obvious problems, such as night, cloudy days, and air pollution that screens out useful rays. How do we carry over these sunless periods? We can't completely — not yet. So when the sun doesn't shine, we'll have to fall back on our old sources of energy, at least until fact and science fiction can be sorted out and some truly workable (and affordable) technology becomes a reality.

Even so, there has been a lot of excitement about experimental solar homes in Colorado, Delaware, Washington, D.C., New Mexico, California, M.I.T., and places abroad. You may have seen the stories. These courageous experiments hold great promise for our collective future.

At this writing a "complete" solar home may be beyond the financial reach of most families in the United States, and there are only a handful of builders around who can tackle such a construction job. But everybody's learning fast, out of necessity, and more complete know-how is not that far down the road.

"Plan ahead," as the sign says. If you're going to build a new home, design it to be flexible, so you can take advantage of some of the new solar technology that will be available soon. There are already lots of options open to you, but there are also some inherent disadvantages and limitations that go along with each type of system.

Give some thought to simple "passive" solar systems as well as the more complicated mechanized ones. (A thick concrete wall, for example, if it's strategically placed, just *sits* there, collecting and storing solar heat during the day and giving it off to the living space in the house at night.) There's a wealth of helpful written material and practical hardware available right now, as the appendix shows. Probably the best book on the subject of solar home design is Donald Watson's best-selling *Designing and Building a Solar House*, also published by Garden Way. We recommend you get a copy.

94

Appendix

The following is a list of manufacturers that have moved into the solar energy field as of October, 1977. As the solar market grows, this list will become less and less complete. For updates, more details on each of these components and/or systems, and for information about special features and options, installation requirements, maintenance requirements, guarantees, warranties, manufacturers' technical services, availability, regional applicability and prices, consult the *Solar Age Catalog: A Guide to Solar Energy Knowledge and Materials.* It's published by SolarVision, Inc., P.O. Box 305, Dover, NJ 07801. The *Solar Age Catalog*'s first edition, which appeared in September, 1977, is done by the same people who publish *Solar Age* magazine, perhaps the finest periodical dealing with practical application of solar energy. The price of their catalog is $8.50.

DuPont Co.
Wilmington, DL 19898
R.C. Robbins; phone: 302/999-3456
Product Name: Teflon FEP Fluorocarbon Film

Heliopticon Corp.
P.O. Drawer 330
Plymouth, NH 03264
David Thun; phone: 603/536-2550
Product Name: Raybend Concentrating Prismatic Sheet

Kalwall Corp.
1111 Candia Road
Manchester, NH 03103
Thomas Minnon; phone: 603/668-8186
Product Names: Sun-Lite Premium; Sun-Lite Glazing Panels; Sunwall

Martin Processing, Inc.
P.O. Box 5068
Martinsville, VA 24112
Raymond A. Woodly; phone: 703/629-1711
Product Name: UX-V Polyester Film

Rohm and Hass Co.
Independence Hall West
Philadelphia, PA 19105
D.T. Espenshade; phone: 215/592-3000
Product Names: Tuffak, Plastic Polycarbonate Sheet and Film; Plexiglas G, Plastic Acrylic Sheet

3M Company
3M Center, Bldg. 223-2
St. Paul, MN 55101
D.H. Flentje; phone: 612/733-2184
Product Name: Flexigard

APPENDIX A
Glazing Manufacturers

A.S.G. Industries, Inc.
P.O. Box 929
Kingsport, TN 37900
Phone: 615/245-0211
Product Name: Sundex and Lo-Iron Glass

CY/RO Industries
Wayne, NJ 07470
Phone: 201/839-4800
Product Name: Acrylite SDP

APPENDIX B
Pipe Insulation Manufacturers

Armaflex
Armstrong Cork Co.
Park 80, Plazza W.
Saddlebrook, NJ 07662
Phone: 201/843-0300
Product Name: Armaflex Pipe Insulation

Owens Corning Fiberglas Corporation
18 Midland St.
Windsor, CT 06095

Michael Smith; phone: 203/249-9358
Product Name: Fiberglas Pipe Insulation

Teledyne Mono-Thane
1460 Industrial Parkway
Akron, OH 44310
Phone: 216/633-6100
Product Name: Foamedge Pipe Cover

Urethane Molding, Inc.
RFD #3 Rt. 11
Lanconia, NH 03246
James M. Annis; phone: 603/524-7577
Product Name: SR6421 Pipe Insulation

APPENDIX C
Absorber Plate Manufacturers

Anaconda Company, Brass Div.
414 Meadow St.
Waterbury, CT 06720
B.W. Erk, Jr.; phone: 203/574-8500
Product Name: Copper Products

Barker Brothers
207 Cortez Ave.
Davis, CA 95616
David Springer; phone: 916/756-4558
Product Name: Absorber Plate

Berry Solar Products
Woodbridge at Main
P.O. Box 327
Edicon, NJ 08817
Calvin C. Beatty; phone: 201/549-3800
Product Name: Absorber Panels

Burton Industries, Inc.
243 Wyandanch Ave.
North Babylon, NY 11704
Burton Z. Chertol; phone: 516/643-6660
Product Name: Sunstrip

Ilse Engineering, Inc.
7177 Arrowhead Rd.
Duluth, MN 55811
John F. Ilse; phone: 218/729-6858
Product Name: Sandwich Panel

Kennecott Copper Corp.
128 Spring St.
Lexington, MA 02173
Phone: 617/862-8268
Product Name: Terra-Light Solar Absorber Plate

Olin Brass
Roll-Bond Products
East Alton, IL 62024
John I. Barton; phone: 618/258-2443
Product Name: Roll-Bond Absorber Plates

Solar Development, Inc.
4180 Westroads Dr.
West Palm Beach, FL 33401
Don Kazimir; phone: 305/842-8935
Product Name: Copper Absorber Plate

The Solarray Corp.
2414 Makiki Hgts. Dr.
Honolulu, Hawaii 96822
Lawrence Judd; phone: 808/533-6464
Product Name: Absorber Plates/ C-18 and C-6

Tranter, Inc.
735 Hazel St.
Lansing, Mich. 48912
Robert S. Rowland; phone: 517/372-8410
Product Name: Solar Absorber Plate/3482S

APPENDIX D
Heat-Transfer Fluid Manufacturers

Dow Corning Corp.
Solar Energy Div.
Midland, Mich. 48640
Richard H. Montgomery; phone: 517/496-4000
Product Name: Q2-1132 Silicone Heat Transfer
    Liquid

Nuclear Technology Corp.
P.O. Box 1
Amston, CT 06231
Thomas F. D'Muhala; phone: 203/537-2387
Product Names: Nutek-800; Nutek-805; Nutek-830;
    Nutek-835; Nutek-876

Sunworks, Division of Enthone, Inc.
P.O. Box 1004
New Haven, CT 06508
Floyd C. Perry, Jr.; phone: 203/934-6301
Product Name: Sunsol 60

APPENDIX E
Storage Tank Manufacturers

Ford Products Corp.
Ford Products Rd.
Valley Cottage, NY 10989
William M. Morrison; phone: 914/358-8282
Product Name: Aqua-Coil Hot Water Storage/Series TC

The Glass-Lined Water Heater Co.
13000 Athens Ave.
Cleveland, OH 44107
Ralph R. Mendelson; phone: 216/521-1377
Product Name: Water Storage Tanks

A.O. Smith Corp.
P.O. Box 28
Kankakee, IL 60901
J.A. Cousins; phone: 815/933-8241
Product Name: Storage Tanks/STJ-30 to 120

APPENDIX F
Liquid Flat Plate Collector Manufacturers

Acorn Structures, Inc.
P.O. Box 250
Concord, MA 01742
Stephen V. Santoro; phone: 617/369-4111
Product Name: Sunwave 420 Solar Collector

All Sunpower, Inc.
10400 S.W. 187th St.
Miami, FL 33157
E.M. Kramer; phone: 305/233-2224
Product Name: All Sunpower/3784

Alten Corporation
2594 Leghorn St.
Mountain View, CA 94043
Klaus Heinmann; phone: 415/969-6474
Product Name: Solar Collector Panels/200G

Ametek, Inc., Power Systems Group
1 Spring Ave.
Hatfield, PA 19440
Frank W. Gilleland; phone: 215/822-2971
Product Name: Copper Solar Collector

Burton Industries, Inc.
243 Wyandanch Ave.
North Babylon, NY 11704
Burton Z. Chertok; phone: 516/643-6660
Product Name: Burton Solar Collector

Chamberlan Manufacturing Corp.
R & D Div. P.O. Box 2545
East 4th and Esther Streets
Waterloo, Iowa 50705
William H. Sima; phone: 319/232-6541
Product Name: Medium Temperature Solar Collector

Cole Solar Systems
440 A East St. Elmo Rd.
Austin, TX 78745
Warren Cole; phone: 512/444-2565
Product Name: Cole Solar Collector/410AT or 410A

Columbia Chase Solar Energy Div.
55 High St.
Holbrook, MA 02343
Walter H. Barret; phone: 617/767-0513
Product Name: Columbia Redi-Mount Collector/77-3376 and 76-3496

Daystar Corp.
90 Cambridge St.
Burlington, MA 01803
Paul P. Chaset; phone: 617/272-8460
Product Name: Daystar '20'

Delta T. Company
2522 West Holly St.
Phoenix, AZ 85009
James L. Hoyer; phone: 602/272-6551
Product Name: Delta T/Model 15

El Camino Solar Systems
5330 Debbie Lane
Santa Barbara, CA 93111
Allen K. Cooper; phone: 805/964-8676
Product Name: Sunspot Solar Collector

Energy Converters, Inc.
2501 N. Orchard Knob Ave.
Chattanooga, TN 37406
David Burrows; phone: 615/624-2608
Product Name: Series 200 Solar Collectors

Gulf Thermal Corp.
P.O. Box 13124
Airgate Branch
Sarasota, FL 33578
Dudley Slocum; phone: 813/355-9783
Product Name: Solar Panel/CUS30 and CUP30

Halstead and Mitchell
P.O. Box 1110
Scottsboro, AL 35768
Product Name: Sunceiver/35775

Lennox Industries, Inc.
P.O. Box 250
200 South 12th Ave.
Marshalltown, IO 50158
Phone: 515/754-4011
Product Name: Lennox Module/LSC18

Libbey-Owens Ford Glass Co.
1701 E. Broadway St.
Toledo, OH 43605
Marty Wenzler; phone: 419/247-4350
Product Name: Lof Sunpanel

Mark M. Manufacturing
RD #2 Box 250
Rexford, NY 12148
Mark Urbaetis; phone: 518/371-9596
Product Name: Mark M Flat Plate Collector

Northrup, Inc.
203 Nichols Dr.
Hutchins, TX 75141
E.C. Ricker; phone: 214/225-4291
Product Name: The Northrup Flat Plate Solar
   Collector/NSC-FPIT

O.E.M. Products, Inc.
Solarmatic Div.
220 W. Brandon Blvd.
Brandon, FL 33511
Product Name: Flat Plate Collector

Owens-Illinois, Inc.
P.O. Box 1035
Toledo, OH 43666
Richard E. Ford; phone: 419/243-1015
Product Name: Sunpak Solar Collector

PAC, A Division of People/Space Co.
49 Garden St.
Boston, MA 02114
Robert Shannon; phone: 617/742-8652
Product Name: PAC Open Flow Collector

Payne, Inc.
1910 Forest Dr.
Annapolis, MD 21401
Fred W. Hawker; phone: 301/268-6150
Product Name: Flat Plate Collector

PPG Industries, Inc.
1 Gateway Center
Pittsburgh, PA 15222
Phone: 412/423-3555
Product Name: Flat Plate Solar Collectors

Revere Copper and Brass, Inc.
Solar Energy Dept.
P.O. Box 151
Rome, NY 13440
William J. Heidrich; phone: 315/338-2401
Product Names: Sunaid; Sun Roof

Reynolds Metals Co.
Torrance Extrusion Plant
2315 Dominguez St.
Torrance, CA 90508
Product Name: Reynolds Solar Collector

Solar Corporation of America
100 Main St.
Warrenton, VA 22186
Walter Sutton; phone: 703/347-7900
Product Names: Mark V Hydronic Collector; Mark III
   Hydronic Collector

Solar Development, Inc.
4180 Westroads Dr.
West Palm Beach, FL 33407
Don Kazimir; phone: 305/842-8935
Product Name: Standard Solar Collector/SD-5

Solar Energy Products, Inc.
1208 N.W. 8th Ave.
Gainesville, FL 32601
Product Name: SEP Flat-Plate Collector, CU30

Solar Energy Research Corp.
701B South Main St.
Longmont, CO 80501
James B. Wiegand; phone: 303/772-8406
Product Name: SERC Liquid Collector Module

Solargenics, Inc.
9713 Lurline Ave.
Chatsworth, CA 91311
Rowen Collins; phone: 213/998-0806
Product Name: Solargenics Collector/Series 76-77

Solargizer Corp.
220 Mulberry St.
Stillwater, MN 55082
William Olson; phone: 612/439-5734
Product Name: Solargizer Solar Collector

Solar Home Systems, Inc.
12931 West Geauga Trail
Chesterland, OH 44026
Joseph Barbish; phone: 216/729-9350
Product Name: Flat Plate Collector SHS/00L1

Solar Innovations
412 Longfellow Blvd.
Lakeland, FL 33801
Ron Yachabach; phone: 813/688-8373
Product Name: Solar Collector Panels/SC-200

SolarKit of Florida, Inc.
1102 139th Ave.
Tampa, FL 33612
Wm. Denver Jones; phone: 813/971-3934
Product Name: Solar Collector

The Solarray Corp.
2414 Makiki Hgts. Dr.
Honolulu, Hawaii 96288
Lawrence Judd; phone: 808/533-6464
Product Name: Flat Plate Collectors/AP-18 and AP-6

Southern Lighting Manufacturing
501 Elwell Ave.
Orlando, FL 32803
Kevin J. Drew; phone: 305/894-8851
Product Name: Universal 100 Thermotube, 45BG &
     410 BG

Sunearth Solar Products Corp.
RD #1 Box 337
Green Lane, PA 18054
H. Katz; phone: 215/699-7892
Product Name: Sunearth Solar Collector/3296 and
     3597A

Sunstream, A Division of Grumman Houston Corp.
P.O. Box 365
Bethpage, NY 11714
Product Name: Sunstream Solar Collector/50

Sunworks, Division of Enthrone, Inc.
P.O. Box 1004
New Haven, CT 06508
Floyd C. Perry, Jr.; phone: 203/934-6301
Product Name: Solector

Unit Electric Control, Inc./Sol-Ray Division
130 Atlantic Dr.
Maitland, FL 32751
Maurice S. Stewart; phone: 305/831-1900
Product Name: Sol-Ray Solar Collector

APPENDIX G
Control System Manufacturers

AeroDesign Co.
P.O. Box 246
Alburtis, PA 18011
R.G. Flower; phone: 215/967-5420
Product Name: Solar Controls/AD-101 and AD-102

Aqueduct Component Group
1537 Pontius Ave.
Los Angeles, CA 90025
Herman Wertheimer; phone: 213/479-3911
Product Name: Automatic Flow Control
     Valve/UP-050 through UP-200

Contemporary Systems, Inc.
68 Charlonne St.
Jaffrey, NH 03452
John Christopher; phone: 603/532-7972
Product Name: LCU-110 Digital Logic Control Unit

Deko-Labs
Box 12841
University Station
Gainesville, FL 32604
Donald F. Dekold; phone: 904/372-6009
Product Name: Solar Differential Controllers

del sol Control Corp.
11914 U.S. #1
Juno, FL 33408
Rodney E. Boyd; phone: 305/626-6116
Product Name: Controls/02A

Energy Applications, Inc.
830 Margie Dr.
Titusville, FL 32780
Napoleon Salvail; phone: 305/269-4893
Product Name: Sun*Trac 100

Energy Converters, Inc.
2501 N. Orchard Knob Ave.
Chattanooga, TN 37406
David Burrows; phone: 615/624-2608
Product Name: Differential Thermostat/E10

Fafco Incorporated
235 Constitution Dr.
Menlo Park, CA 94025
Phone: 415/321-3650
Product Name: Fafco Solar Control

Hawthorne Industries, Inc.
1501 Dixie Highway
West Palm Beach, FL 33401
Ray Lewis; phone: 305/659-5400
Product Name: Fixflo and Variflo Controllers/H-1503,
    H-1515, H-1510, H-1512

Heliotrope General
3733 Kenora Dr.
Spring Valley, CA 90277
Sam Dawson; phone: 714/460-3930
Product Name: Delta-T/DTT-70 through ATT-3414

Honeywell, Inc.
Residential Division Customer Service
1885 Douglas Drive North

Minneapolis, MN 55422
Phone: 612/542-7500
Product Name: Solar Temperature Control/R7406A

Natural Power, Inc.
New Boston, NY 03070
Rick Katzenberg; phone: 603/487-2426
Product Name: Solar Control/SC100

Rho Sigma, Inc.
15150 Raymer St.
Van Nuys, CA 94105
Robert Schlesinger; phone: 213/342-4376
Product Names: Differential Thermostat with Freeze
    Protection/RS 104; Differential Thermostat/RS
    106; Proportional Control with Dual Output/RS
    500P Series; Differential Thermostat/RS #240

Robertshaw Controls Co.
100 W. Victoria
Long Beach, CA 90805
Preston Welch; phone: 213/638-6111
Product Name: Solar Commander/SD-10

Simons Solar Environmental Systems
24 Carlisle Pike
Mechanicsburg, PA 17055
C. John Urling; phone: 717/697-2778
Product Name: Simons Solar Differential
    Thermostats

Solar Control Corp.
5595 Arapahoe Rd.
Boulder, CO 80302
Thomas B. Kent; phone: 303/449-9180
Product Names: Solar Heating and Cooling Controls;
    Differential Thermostats

Solar Energy Research Corp.
701B South Main St.
Longmont, CO 80501
James B. Wiegand; phone: 303/772-8406
Product Name: Thermo-Mate Deluxe Controller,
    DC-761 and Standard Controller, SC-762

Solarics
P.O. Box 15183
Plantation, FL 33318
Ronald Stein
Product Name: Solid State Differential
    Controller/SPC-2000 Series, SPC-1000 Series

West Wind Co.
Box 1465
Farmington, NM 87401
Geoffrey Gerhard; phone: 505/325-4949
Product Name: Temperature Differential Switch

APPENDIX H
Circulation Pump Manufacturers

Grundfos Pump Corp.
2555 Clovis Ave.
Clovis, CA 93612
Evelyn Graham; phone: 209/299-9741
Product Names: Pump, Hot Water Circulator/UMS
   20-28; UP 25-42 SF; UPS 20-42 F; UP 26-64 F;
   Pump, Multi-Stage/CP 2,3,8; CR 30

Solar Innovations
412 Longfellow Blvd.
Lakeland, FL 33801
Ron Yachabach; phone: 813/688-8373
Product Name: Electronically Controlled Circulating
   Pump System/PA-201

APPENDIX I
Complete Solar Hot Water Systems

Cole Solar Systems
440A East St. Elmos Rd.
Austin, TX 78745
Warren Cole; phone: 512/444-2565
System Name: Domestic Hot Water System/12A or
   28D

Columbia Chase Solar Energy Div.
55 High St.
Holbrook, MA 02343
Walter H. Barret; phone: 617/767-0513
System Names: Columbia Domestic Hot Water
   System; Columbia Direct Exchange Domestic Hot
   Water Kit

El Camino Solar Systems
5330 Debbie Lane
Santa Barbara, CA 93111

Allen K. Cooper; phone: 805/964-8676
System Name: Sunspot Solar Water Heating System

Energy Applications, Inc.
830 Margie Dr.
Titusville, FL 32780
Napoleon P. Salvail; phone: 305/269-4893
System Name: Sun-tracking Residential Solar Hot
   Water System/1000

Energy Converters, Inc.
2501 N. Orchard Knob Ave.
Chattanooga, TN 37406
David Burrows; phone: 615/624-2608
System Name: Solar Saver/SWH801

Energy Converters, Inc.
2501 N. Orchard Knob Ave.
Chattanooga, TN 37406
David Burrows; phone: 615/624-2608
System Name: Suntrap/SWH401

Groundstar Energy Corp.
137 Rowayton Ave.
Rowayton, CT 06853
System Name: Groundstar Domestic Hot Water
   System

Heilmann Electric
127 Mountainview Rd.
Warren, NJ 07060
Phone: 201/757-4507
System Name: Solar Tube Systems Water Heater Kits

W.L. Jackson Mfg. Co., Inc.
1200-26 E. 40th St.
P.O. Box 11168
Chattanooga, TN 37401
Ralph L. Braly; phone: 615/867-4700
System Name: Water Heating System

Mor-Flo Industries, Inc.
18450 S. Miles Rd.
Cleveland, OH 44128
Phone: 216/663-7300
System Name: Solarstream Domestic Hot Water
   System

Northrup, Inc.
302 Nichols Dr.
Hutchins, TX 75141
E.C. Ricker; phone: 214/225-4291
System Name: The Northrup
   Thermosiphon/NSC-TSWH

Raypak, Inc.
31111 Agoura Rd.
Westlake Village, CA 91361
H. Byers or A. Boniface; phone: 213/889-1500
System Name: Raypak/DHWS-2-T82

Revere Copper and Brass, Inc.
Solar Energy Dept.
P.O. Box 151
Rome, NY 13440
William J. Heidrich; phone: 315/338-2401
System Name: Solar Domestic Water Heating System

Solar Development, Inc.
4180 Westroads Dr.
West Palm Beach, FL 33407
Don Kazimir; phone: 305/842-8935
System Name: Solar Water Heater/SD-5

Solar Energy Corp.
701B South Main St.
Longmont, CO 80501
James B. Wiegand; phone: 303/772-8406
System Name: SERC Domestic Hot Water Solar
   Preheater

Solargenics, Inc.
9713 Lurline Ave.
Chatsworth, CA 91311
Rowen Collins; phone: 213/998-0806
System Name: Sol-Pak IV; Residential Hot Water
   Systems/SP 66 through SP 120

SolarKit of Florida
1102 139th Ave.
Tampa, FL 33612
Wm. Denver Jones; phone: 813/971-3934
System Name: Solar Hot-Water Kit

Solaron Corporation
300 Galleria Tower
720 S. Colorado Blvd.
Denver, CO 80222
System Name: Hot Water Systems

State Industries, Inc.
Ashland City, TN 37015
or
Henderson, NV 89015
System Name: Solarcraft/1 & 11

Sunstream, Division of Grumman Houston Corp.
P.O. Box 365
Bethpage, NY 11714

System Name: Sunstream Domestic Hot Water
   System/50

Sunearth Solar Products Corp.
RD 1 Box 337
Green Lane, PA 18054
H. Katz; phone: 215/699-7892
System Name: Sunearth Domestic Hot Water
   Systems

Sunrise Solar Services
1174 Blossom St.
Suffield, CT 06078
Rick Schwolsky; phone: 203/668-0349
System Name: Solar Water Heating Systems

Sunworks, Division of Entrone, Inc.
P.O. Box 1004
New Haven, CT 06508
Floyd C. Perry, Jr.; phone: 203/934-6301
System Name: Solector Pak 1000 Series; Solector
   Pak 3000 Series

The Wilcon Corporation
3310 S.W. 7th St.
Ocala, FL 32670
Jack W. Rainford; phone: 904/732-2550
System Name: Wilcon Solar Water Heating
   System/120-6

APPENDIX J
Solar Swimming Pool Heaters

Aquasolar, Inc.
1232 Zacchini Ave.
Sarasota, FL 33577
G.J. Zella or John Pickett; phone: 813/336-7080
System Name: Aquasolar

Burke Industries, Inc.
2250 S. Tenth St.
San Jose, CA 95112
Larry R. Schader; phone: 408/297-3500
System Name: Burke Solar Heater

Cole Solar Systems
440A East St. Elmo Rd.
Austin, TX 78745
Warren Cole; phone: 512/444-2565
System Name: Swimming Pool Collector

Fafco, Inc.
138 Jefferson Dr.
Menlo Park, CA 94025
Larry Hix; phone: 415/321-6311
System Name: Swimming Pool Heating System

H.C. Products Co.
P.O. Box 68
Princeville, IL 61559
Wm. Foster; phone: 412/553-3185
System Name: ALCOA Solar Heating Systems for
    Swimming Pools

Raypak, Inc.
31111 Agoura Rd.
Westlake Village, CA 91361
R. Dominquez or A. Boniface; phone: 213/889-1500
System Name: Solar-Pak/SK800 and SK1000

Solar Development, Inc.
4180 Westroads Dr.
West Palm Beach, FL 33407
Don Kazimir; phone: 305/842-8935
System Name: SDI Swimming Pool Heater

Solar Home Systems, Inc.
12931 West Geauga Trail
Chesterland, OH 44026
Joseph Barbish; phone: 216/729-9350
System Name: Solar Pool Heater/SPH-001

APPENDIX K
Manufacturers or Distributors
of Solenoid Valves

Heliotrope General
3733 Kenora Dr.
Spring Valley, Ca 90277
Sam Dawson; phone: 714/460-3930
Product Name: Solenoid Valves

Richdel, Inc.
1851 Oregon St.
Carson City, NV 89701
Dale Soukup; phone: 702/882-6786
Product Name: Electric Solenoid Valve

Bibliography
and
Suggested Reading

Allred, Johnny W., et al. *An Inexpensive Economical Solar Heating System for Homes.* Langley Research Center, Hampton, Va., 1976.

A.I.A. Research Corp. *Solar Dwelling Design Concepts.* H.U.D., Washington, D.C., 1976.

"Alternative Technology Equipment Directory." *Spectrum,* Alternative Sources of Energy, Milaca, Minn., 1975.

Baer, Steve. *Solar Water Heater.* Zomeworks Corp., Albuquerque, N.M., 1974.

Baer, Steve. *Sunspots.* Zomeworks Corp., Albuquerque, N.M., 1975.

Beste, Frederick and Beste, Virginia. *Free Power.* Consolidated Energy Consultants, Inc., Baltimore. 1975.

Branley, Franklyn M. *Solar Energy.* Thomas Y. Crowell, N.Y., 1975.

Brinkworth, B.J. *Solar Energy for Man.* John Wiley and Sons, N.Y., 1973.

Clark, Wilson. *Energy for Survival.* Anchor Press/Doubleday, Garden City, N.Y., 1975.

Converse, A.O. *The Assessment of Solar-Heated Buildings and Panels.* Dartmouth College, Hanover, N.H., 1975.

Daniels, Farrington. *Direct Uses of the Sun's Energy.* Yale University Press, New Haven, Conn., 1964.

Daniels, George. *Home Guide to Plumbing, Heating, Air Conditioning.* Popular Science Publishing Co., N.Y., 1967.

Daniels, George. *Solar Homes and Sun Heating.* Harper and Row, N.Y., 1976.

DeVries, John. *Sol-R-Tech Operations Manual.* Sol-R-Tech, Hartford, Vt., 1975.

Energy for Rural Development. National Academy of Sciences, Washington, D.C., 1976.

Energy Primer. Portola Institute, Menlo Park, Cal., 1974.

Foster, William M. *Homeowners Guide to Solar Heating and Cooling.* TAB Books, Blue Ridge Summit, Pa., 1976.

Keyes, John. *Harnessing the Sun.* Morgan and Morgan, Dobbs Ferry, N.Y., 1974.

Handbook of Homemade Power, Bantam Books, N.Y., 1974.

Hoke, John. *Solar Energy.* Franklin Watts, N.Y., 1976.

How to Build a Solar Water Heater. Florida Conservation Foundation, Inc., Winter Park, Fla., 1975.

Lucas, Ted. *How to Build a Solar Heater.* Ward Ritchie Press, Pasadena, Cal., 1975.

Noll, Edward M. *Wind/Solar Energy.* Bobbs-Merrill, N.Y., 1975.

Rau, Hans. *Solar Energy.* Macmillan, N.Y., 1964.

Skarbek, Habdank. *Save Heating Costs: Use Solar Energy.* Keystone Solar Energy, Inc., N.Y., 1975.

Solar Age Catalog. Dover, N.J., 1977.

"Solar Collector Installation." *Workbench,* Modern Handcraft Inc., Kansas City. MO., vol. 33, number 3, 1977.

Solar Domestic Water Heating. Sunworks, Inc., Guilford, Conn., 1975.

"Solar Swimming Season." *Solar Age,* Dover, N.J., vol. 2, number 3, 1977.

Stoner, Carol, et al. *Producing Your Own Power.* Rodale Press, Emmaus, Pa., 1974.

Watson, Donald. *Designing and Building a Solar House.* Garden Way Publishing, Charlotte, Vt., 1977.

Whitehouse, Harry, et al. *Other Homes and Garbage.* Sierra Club Books, San Francisco, 1975.

Williams, J. Richard. *Solar Energy.* Ann Arbor Science, Ann Arbor, Mich., 1975.

Index

Other Garden Way Books You Will Enjoy

The own/builder and the home-owner concerned about energy conservation and alternate construction methods will find an up-to-date library essential. Here are some excellent books in these areas from the publisher of **Build Your Own Solar Water Heater.**

Harnessing Water Power for Home Energy, by Dermot McGuigan. 112 pp., quality paperback, $4.95, library hardcover, $9.95. An authoritative, detailed look at the uses of water power for small-scale operations.

Harnessing Wind Power for Home Energy, by Dermot McGuigan. 144 pp., quality paperback, $4.95, library hardcover, $9.95. A thorough review of wind power systems for the homeowner, with illustrations and manufacturers' directory.

Low-Cost Pole Building Construction, by Douglas Merrilees and Evelyn Loveday. 118 pp., deluxe paperback, $5.95. This will save you money, labor, time and materials.

Build Your Own Stone House, by Karl and Sue Schwenke. 156 pp., quality paperback, $5.95; hardback, $8.95. With their help, you can build your own beautiful stone home.

New Low-Cost Sources of Energy for the Home, by Peter Clegg. 250 pp., quality paperback, $7.95; hardback, $10.95. Covers solar heating and cooling, wind and water power, wood heat and methane digestion. Packed with information.

The Complete Book of Heating with Wood, by Larry Gay. 128 pp., quality paperback, $3.95. Fight rising home heating costs and still keep very warm.

Wood Stove Know-how, by Peter Coleman. 24 pp., illustrated paperback, $1.50. Installation, cleaning and maintenance instructions, plus much more.

The Complete Homesteading Book, by David Robinson, 256 pp., quality paperback, $5.95; hardback, $9.95. How to live a simpler, more self-sufficient life.

Buying Country Property, by Herb Moral. 128 pp., quality paperback, $3.95. Sure to be your "best friend" when considering country property.

Build Your Own Log Home, by Roger Hard. 204 pp., quality paperback, $6.95; hardback, $10.95. A real guidebook to building with logs or log home kits, by a man who has built his own log home and others.

Designing & Building a Solar House, by Donald Watson. 288 pp., quality paperback, $8.95; hardback, $12.95. The best and most thorough book yet on solar houses, with over 400 illustrations.

Your Energy-Efficient House, by Anthony Adams. 128 pp., quality paperback, $4.95. The perfect idea book for those concerned about saving energy in a new or existing house.

Building & Using Our Sun-Heated Greenhouse, by Helen & Scott Nearing. 156 pp., quality paperback, $6.95; hardback, $9.95. The Nearings share, in text and photographs, the secrets they have learned in over 50 years of gardening in New England all year-round.

Methanol & Other Ways Around the Gas Pump, by John Ware Lincoln. 144 pages; quality paperback, $4.95. How to "drive without gas" — using methanol — and a look at the past experiments and future politics of our gasoline supply.

These Garden Way books are available at your bookstore, or may be ordered directly from Garden Way Publishing, Dept. 171X, Charlotte, Vermont 05445. If your order is less than $10, please add 60¢ postage and handling.